VOLUME EIGHTY FIVE

ADVANCES IN PARASITOLOGY

SERIES EDITOR

EDITORIAL BOARD

VOLUME EIGHTY FIVE

ADVANCES IN PARASITOLOGY

Edited by

D. ROLLINSON

Life Sciences Department
The Natural History Museum
London, UK

J. R. STOTHARD

Department of Parasitology
Liverpool School of Tropical Medicine
Liverpool, UK

AMSTERDAM • BOSTON • HEIDELBERG • LONDON
NEW YORK • OXFORD • PARIS • SAN DIEGO
SAN FRANCISCO • SINGAPORE • SYDNEY • TOKYO
Academic Press is an imprint of Elsevier

Academic Press is an imprint of Elsevier

32 Jamestown Road, London, NW1 7BY, UK
525 B Street, Suite 1800, San Diego, CA 92101-4495, USA
225 Wyman Street, Waltham, MA 02451, USA
Radarweg 29, PO Box 211, 1000 AE Amsterdam, The Netherlands

First edition 2014

Notice

No responsibility is assumed by the publisher for any injury and/or damage to persons or property as a matter of products liability, negligence or otherwise, or from any use or operation of any methods, products, instructions or ideas contained in the material herein. Because of rapid advances in the medical sciences, in particular, independent verification of diagnoses and drug dosages should be made.

ISBN: 978-0-12-800182-0
ISSN: 0065-308X

For information on all Academic Press publications visit our website at store.elsevier.com

CONTENTS

CONTRIBUTORS

Cristian A. Alvarez Rojas
Centre for Animal Biotechnology, Faculty of Veterinary Science, The University of Melbourne, Parkville, Victoria, Australia

Cova R. Arias
Aquatic Microbiology Laboratory, School of Fisheries, Aquaculture and Aquatic Sciences, College of Agriculture, Auburn University, Auburn, Alabama, USA

Stephen A. Bullard
Aquatic Parasitology Laboratory, School of Fisheries, Aquaculture and Aquatic Sciences, College of Agriculture, Auburn University, Auburn, Alabama, USA

Thomas H. Cribb
School of Biological Sciences, The University of Queensland, Brisbane, Queensland, Australia

Robin B. Gasser
Faculty of Veterinary Science, The University of Melbourne, Parkville, Victoria, Australia

Kenneth M. Halanych
Department of Biological Sciences, Auburn University, Auburn, Alabama, USA

Aaron R. Jex
Faculty of Veterinary Science, The University of Melbourne, Parkville, Victoria, Australia

Alan J. Lymbery
Parasitology, and Freshwater Fish Group & Fish Health Unit, School of Veterinary and Life Sciences, Murdoch University, Perth, Western Australia, Australia

Raphael Orélis-Ribeiro
Aquatic Parasitology Laboratory, School of Fisheries, Aquaculture and Aquatic Sciences, College of Agriculture, Auburn University, Auburn, Alabama, USA

Jean-Pierre Y. Scheerlinck
Centre for Animal Biotechnology, Faculty of Veterinary Science, The University of Melbourne, Parkville, Victoria, Australia

R.C. Andrew Thompson
Parasitology, School of Veterinary and Life Sciences, Murdoch University, Perth, Western Australia, Australia

Amanda R. Worth
Parasitology, School of Veterinary and Life Sciences, Murdoch University, Perth, Western Australia, Australia

PREFACE

Professor Russell Stothard, Medical Parasitologist at the Liverpool School of Tropical Medicine, joins as Co-Editor of *Advances of Parasitology* after serving for several years on the Editorial Boards of *Transactions of the Royal Society of Tropical Medicine and Hygiene*, *International Health* and more recently *Parasitology*. Russell was also Editor of the *Bulletin of International Health and Tropical Medicine*, the quarterly newsletter for the *RSTMH* and is now general secretary for the British Society for Parasitology. His main research has been as a medical parasitologist, but Russ also has more general interests in the field of parasitology that spans from ecology to epidemiology including both helminths and protists and as a keen angler he includes some fish parasitology dispersed in between. Russ looks forward to joining David in this important role to ensure that *Advances* continues to be the leading place for expert reviews which maintain the vibrancy and impact of this important discipline within the fields of medical, veterinary and life sciences studies.

CHAPTER ONE

Diversity and Ancestry of Flatworms Infecting Blood of Nontetrapod Craniates "Fishes"

Raphael Orélis-Ribeiro*, Cova R. Arias†, Kenneth M. Halanych‡, Thomas H. Cribb§, Stephen A. Bullard*,1

*Aquatic Parasitology Laboratory, School of Fisheries, Aquaculture and Aquatic Sciences, College of Agriculture, Auburn University, Auburn, Alabama, USA
†Aquatic Microbiology Laboratory, School of Fisheries, Aquaculture and Aquatic Sciences, College of Agriculture, Auburn University, Auburn, Alabama, USA
‡Department of Biological Sciences, Auburn University, Auburn, Alabama, USA
§School of Biological Sciences, The University of Queensland, Brisbane, Queensland, Australia
[1]Corresponding author: e-mail address: sab0019@auburn.edu

Contents

Abstract

We herein review all published molecular studies (life history, taxonomy, and phylogeny) and summarize all GenBank sequences and primer sets for the "fish blood flukes". Further, by analysing new and all available sequence data for the partial D1–D2 domains of 28S from 83 blood fluke taxa, we explore the evolutionary expansion of flatworm parasitism in the blood of craniates. Based on this analysis, the blood flukes infecting marine bony fishes (Euteleostei) are monophyletic. The clade comprising the chondrichthyan blood fluke plus the marine euteleost blood flukes is the sister group to tetrapod blood flukes (spirorchiids and schistosomes). The innominate blood fluke cercariae from freshwater gastropods were monophyletic and sister to the clade comprising spirorchiids and schistosomes, but low nodal support indicated that they may represent a distinct blood fluke lineage with phylogenetic affinities also to fish blood flukes. Blood flukes that utilize gastropod intermediate hosts were monophyletic

Advances in Parasitology, Volume 85
ISSN 0065-308X
http://dx.doi.org/10.1016/B978-0-12-800182-0.00001-5

(unidentified gastropod cercariae + tetrapod blood flukes) and those utilizing bivalves and polychaetes were monophyletic (marine fish blood flukes). Low or no taxon sampling among blood flukes of basal fish lineages and primary division freshwater fish lineages are significant data gaps needing closure. We also note that no record of an infection exists in a hagfish (Myxiniformes), lamprey (Petromyzontiformes), or nontetrapod sarcopterygiian, i.e., coelacanth (Coelacanthimorpha) or lungfish (Dipnoi). The present phylogenetic analysis reiterated support for monophyly of Schistosomatidae and paraphyly of spirorchiids, with the blood flukes of freshwater turtles basal to those of marine turtles and schistosomes.

1. INTRODUCTION

Blood flukes (Platyhelminthes: Digenea: Schistosomatoidea) historically have been assigned to three families, each corresponding to the vertebrate definitive host lineages they infect (Amemiya et al., 2013; Nelson, 2006). Fish blood flukes (Digenea: Aporocotylidae; also as "Sanguinicolidae"; hereafter referred to as "FBFs") (Bullard et al., 2009; Smith, 1972, 1997a,b, 2002) infect nontetrapod craniates, i.e., paraphyletic fishes. Turtle blood flukes (Digenea: paraphyletic "Spirorchiidae") principally infect marine and freshwater turtles (Chelonia) (Snyder, 2004), with recent molecular phylogenetic support for inclusion of the crocodile-infecting, dioecious blood fluke *Griphobilharzia amoena* (see Brant and Loker, 2005; Loker and Brant, 2006; Platt et al., 1991, 2013). Schistosomes (Digenea: Schistosomatidae) (see Lockyer et al., 2003b) infect birds and mammals (Cribb et al., 2001; Olson et al., 2003) and have been the most studied, making them among the best known of trematode families (Brant et al., 2006). They cause human schistosomiasis and are among the world's most economically important metazoan parasites, with the species of *Schistosoma* (principally *S. mansoni*, *S. japonicum*, and *S. haematobium*) infecting >230 million people from 28 countries and killing an estimated 280,000 people annually in sub-Saharan Africa (Rollinson et al., 2013; Van der Werf et al., 2003; World Health Organization, 2010). Most phylogenetic attention has focused on schistosomes, and abundant evidence exists that they comprise a monophyletic group. Two molecular phylogenetic studies have been published for turtle blood flukes (Snyder, 2004; Tkach et al., 2009) and nine for those of FBFs. The latter analyses typically include a limited number of taxa and markers (Tables 1.1–1.3) and treat relationships between or within genera, natural history, or life cycles (Aiken et al., 2007; Alama-Bermejo et al., 2011;

Table 1.1 GenBank DNA sequences for fish blood flukes

Parasite	Host	Stage	Locality	Setting	GenBank accession numbers				Reference(s)
					18S	28S	ITS2	COI	
Acipensericola petersoni	*Polyodon spathula*	A	Mississippi Delta, the United States	Wild	DQ534192[a]				Bullard et al. (2008)
Ankistromeces dunwichensis	*Siganus fuscescens*	A	SW Pacific, off North Stradbroke Island, Australia	Wild			DQ335838[b]		Nolan and Cribb (2006a)
Ankistromeces mariae	*Meuschenia freycineti*	A	SW Pacific, off Stanley Harbour, Australia	Wild			DQ335839[b]		Nolan and Cribb (2006a)
Ankistromeces olsoni	*Siganus fuscescens*	A	SW Pacific, off Heron Island, Australia	Wild			DQ335840[b]		Nolan and Cribb (2006a)
Aporocotyle argentinensis	*Merluccius hubbsi*	A	SW Atlantic, off Patagonia, Argentina	Wild		JX094803[c]			Hernández-Orts et al. (2012)
Aporocotyle mariachristinae	*Genypterus blacodes*	A	SW Atlantic, off Patagonia, Argentina	Wild	JX094801[a]	JX094802[c]			Hernández-Orts et al. (2012)

Continued

Table 1.1 GenBank DNA sequences for fish blood flukes—cont'd

Parasite	Host	Stage	Locality	Setting	GenBank accession numbers				Reference(s)
					18S	28S	ITS2	COI	
Aporocotyle spinosicanalis	*Merluccius merluccius*	A	NE Atlantic	Wild		AF167094[c]			Snyder and Loker (2000)
		A	NE Atlantic, the United Kingdom	Wild	AJ287477[b]				Cribb et al. (2001)
		A	NE Atlantic, off Orkney Islands, the United Kingdom	Wild		AY222177[c]			Olson et al. (2003)
Braya jexi	*Scarus frenatus*	A	SW Pacific, off Heron Island, Australia	Wild			DQ059624[b]		Nolan and Cribb (2006b)
Braya psittacus	*Scarus ghobban*	A	SW Pacific, off Heron Island, Australia	Wild			DQ059625[b]		Nolan and Cribb (2006b)
Braya yantschi	*Chlorurus microrhinos*	A	SW Pacific, off Heron Island, Australia	Wild			DQ059628[b]		Nolan and Cribb (2006b)

Cardicola aurata	*Sparus aurata*	A	Mediterranean Sea, off Valencia, Spain	Cultured	AM910616[c]	AM910617[b]	Holzer et al. (2008)
Cardicola bartolii	*Siganus lineatus*	A	SW Pacific, off Heron Island, Australia	Wild		DQ059631[b]	Nolan and Cribb (2006b)
Cardicola chaetodontis	*Chaetodon aureofasciatus*	A	SW Pacific, off Lizard Island, Australia	Wild		KF049000[b]	Yong et al. (2013)
	Chaetodon baronessa	E	SW Pacific, off Lizard Island, Australia	Wild		KF049004[b]	Yong et al. (2013)
	Chaetodon kleinii	E	SW Pacific, off Heron Island, Australia	Wild		JN418931[b]	Yong et al. (2013)
	Chaetodon lunulatus	E	SW Pacific, off Lizard Island, Australia	Wild		KF049002[b]	Yong et al. (2013)
	Chaetodon plebeius	E	SW Pacific, off Lizard Island, Australia	Wild		KF049003[b]	Yong et al. (2013)

Continued

Table 1.1 GenBank DNA sequences for fish blood flukes—cont'd

Parasite	Host	Stage	Locality	Setting	GenBank accession numbers				Reference(s)
					18S	28S	ITS2	COI	
	Chaetodon rainfordi	A	SW Pacific, off Heron Island, Australia	Wild			JN418932[b]		Yong et al. (2013)
	Chaetodon unimaculatus	E	SW Pacific, off Heron Island, Australia	Wild			DQ059633[b]		Nolan and Cribb (2006b)
	Chaetodon unimaculatus	E	SW Pacific, off Lizard Island, Australia	Wild			KF049001[b]		Yong et al. (2013)
Cardicola coeptus	*Siganus punctatus*	A	SW Pacific, off Heron Island, Australia	Wild			DQ059629[b]		Nolan and Cribb (2006b)
		A	SW Pacific, off Heron Island, Australia	Wild		JF803976[c]			Cribb et al. (2011)
Cardicola coeptus	*Siganus vulpinus*	A	SW Pacific, off Heron Island, Australia	Wild			DQ059630[b]		Nolan and Cribb (2006b)
		A	SW Pacific, off Heron Island, Australia	Wild		JF803977[c]			Cribb et al. (2011)

Cardicola covacinae	*Siganus punctatus*	A	SW Pacific, off Heron Island, Australia	Wild		DQ059634[b]	Nolan and Cribb (2006b)
Cardicola currani	*Sciaenops ocellatus*	A	Gulf of Mexico, off Davis Bayou, the United States		KJ272524[c]		Present study
Cardicola forsteri	*Longicarpus modestus*	C	SW Pacific, off Port Lincoln, S Australia	Wild	JF800668[c]	JF800670[b]	Cribb et al. (2011)
Cardicola forsteri	*Thunnus maccoyii*	A	SW Pacific, off S Australia	Cultured		DQ059637[b]	Nolan and Cribb (2006b)
		A	SW Pacific, off Port Lincoln, S Australia	Cultured	EF653387[c]	EF661575[b]	Aiken et al. (2007)
		A	SW Pacific, off Cabbage Patch, Australia	Wild	EF653389[c]	EF653394[b]	Aiken et al. (2007)
		A	SW Pacific, off Port Lincoln, S Australia	Cultured	AB742426[c]	AB742428[b]	Shirakashi et al. (2013)

Continued

Table 1.1 GenBank DNA sequences for fish blood flukes—cont'd

Parasite	Host	Stage	Locality	Setting	GenBank accession numbers				Reference(s)
					18S	28S	ITS2	COI	
Cardicola forsteri	*Thunnus thynnus*	A	Mediterranean Sea, off Puerto de Mazarrón, Spain	Cultured		EF653388[c]	EF653395[b]		Aiken et al. (2007)
Cardicola lafii	*Siganus fuscescens*	A	SW Pacific, off Lizard Island, Australia	Wild			DQ059639[b]		Nolan and Cribb (2006b)
Cardicola milleri	*Lutjanus bohar*	A	SW Pacific, off Lizard Island, Australia	Wild			DQ059640[b]		Nolan and Cribb (2006b)
Cardicola opisthorchis	*Thunnus orientalis*	A	NW Pacific, Japan	Cultured		HQ324227[c]	HQ324228[b]		Ogawa et al. (2011)
Cardicola opisthorchis	*Terebella* sp.	C	NW Pacific, off Tsushima, Japan	Wild		AB829900[c]	AB830082[b]		Sugihara et al. (2014)
Cardicola orientalis	*Thunnus maccoyii*	A	SW Pacific, off Port Lincoln, S Australia	Cultured		AB742425[c]	AB742427[b]		Shirakashi et al. (2013)
Cardicola orientalis	*Thunnus orientalis*	A	NW Pacific, off Kushimoto, Japan	Cultured		HQ324225[c]	HQ324226[b]		Ogawa et al. (2011)

Cardicola palmeri	*Pogonias cromis*	A	Gulf of Mexico, off Back Bay, the United States			KJ572525[c]			Present study
Cardicola parilus	*Siganus fuscescens*	A	Indian, off Ningaloo Reef, W Australia	Wild			DQ059638[b]		Nolan and Cribb (2006b)
Cardicola tantabiddii	*Siganus fuscescens*	A	Indian, off Ningaloo Reef, W Australia	Wild			DQ059642[b]		Nolan and Cribb (2006b)
Cardicola watsonensis	*Siganus corallinus*	A	SW Pacific, off Lizard Island, Australia	Wild			DQ059643[b]		Nolan and Cribb (2006b)
Chimaerohemecus trondheimensis	*Chimaera monstrosa*	A	NE Atlantic, off Bergen, Norway	Wild	AY157213[a]	AY157239[a]		AY157185[a]	Lockyer et al. (2003b)
Elaphrobates euzeti	*Lutjanus campechanus*	A	Gulf of Mexico, USA			KJ572526[c]			Present study
Neoparacardicola nasonis	*Naso unicornis*	A	SW Pacific, off Lizard Island, Australia	Wild	AY222097[a]	AY222179[c]			Olson et al. (2003)

Continued

Table 1.1 GenBank DNA sequences for fish blood flukes—cont'd

Parasite	Host	Stage	Locality	Setting	GenBank accession numbers				Reference(s)
					18S	28S	ITS2	COI	
Paracardicoloides yamagutii	*Anguilla reinhardtii*	A	Churchbank Weir, Australia	Wild	U42569[c]	U42562[c]			Barker and Blair (1996)
		A	Brisbane River tributaries, Australia	Wild			AY465872[b]		Nolan and Cribb (2004a)
	Posticobia brazieri	C	Brisbane River tributaries, Australia	Wild			AY465869[b]		Nolan and Cribb (2004a)
Paradeontacylix balearicus	*Seriola dumerili*	A	Mediterranean Sea, off Majorca, Spain	ns		AM489594[c]	AM489600[b]	AM489604[c]	Repullés-Albelda et al. (2008)
Paradeontacylix godfreyi	*Seriola lalandi*	A	SW Pacific, off Port Lincoln, S Australia	ns		AM489597[c]	AM489602[b]	AM489607[c]	Repullés-Albelda et al. (2008)
Paradeontacylix grandispinus	*Seriola dumerili*	A	NW Pacific, off Ushine, Japan	ns		AM489596[c]	AM489601[b]		Repullés-Albelda et al. (2008)
Paradeontacylix ibericus	*Seriola dumerili*	A	Mediterranean Sea, off Santa Pola, Spain	ns		AM489593[c]	AM489598[b]	AM489603[c]	Repullés-Albelda et al. (2008)

Paradeontacylix kampachi	*Seriola dumerili*	A	NW Pacific, off Ushine, Japan	ns	AM489595[c]	AM489599[b]	AM489605[c], AM489606[c]	Repullés-Albelda et al. (2008)
Pearsonellum corventum	*Plectropomus leopardus*	A	SW Pacific, off Heron Island, Australia	Wild		AY465873[b]		Nolan and Cribb (2004b)
Pearsonellum pygmaeus	*Cromileptes altivelis*	A	SW Pacific, off Lizard Island, Australia	Wild		AY465874[b]		Nolan and Cribb (2004b)
Phthinomita adlardi	*Siganus argenteus*	A	Indian, off Ningaloo Reef, W Australia	Wild		DQ335844[b]		Nolan and Cribb (2006a)
Phthinomita brooksi	*Siganus virgatus*	A	Indian, off Ningaloo Reef, W Australia	Wild		DQ335845[b]		Nolan and Cribb (2006a)
Phthinomita hallae	*Siganus corallinus*	A	SW Pacific, off Heron Island, Australia	Wild		DQ335846[b]		Nolan and Cribb (2006a)
	Siganus doliatus	A	SW Pacific, off Heron Island, Australia	Wild		DQ335847[b]		Nolan and Cribb (2006a)

Continued

Table 1.1 GenBank DNA sequences for fish blood flukes—cont'd

Parasite	Host	Stage	Locality	Setting	GenBank accession numbers				Reference(s)
					18S	28S	ITS2	COI	
	Siganus vulpinus	A	SW Pacific, off Heron Island, Australia	Wild			DQ335848[b]		Nolan and Cribb (2006a)
Phthinomita ingramae	*Siganus punctatus*	A	SW Pacific, off Lizard Island, Australia	Wild			DQ335849[b]		Nolan and Cribb (2006a)
Phthinomita jonesi	*Siganus argenteus*	A	SW Pacific, off Lizard Island, Australia	Wild			DQ335850[b]		Nolan and Cribb (2006a)
	Siganus doliatus	A	SW Pacific, off Lizard Island, Australia	Wild			DQ335851[b]		Nolan and Cribb (2006a)
	Siganus lineatus	A	SW Pacific, off Lizard Island, Australia	Wild			DQ335852[b]		Nolan and Cribb (2006a)
	Siganus vulpinus	A	SW Pacific, off Lizard Island, Australia	Wild			DQ335853[b]		Nolan and Cribb (2006a)
Phthinomita littlewoodi	*Siganus corallinus*	A	SW Pacific, off Lizard Island, Australia	Wild			DQ335854[b]		Nolan and Cribb (2006a)

	Siganus lineatus	A	SW Pacific, off Heron Island, Australia	Wild	DQ335855[b]	Nolan and Cribb (2006a)
Phthinomita munozae	*Choerodon venustus*	A	SW Pacific, off Heron Island, Australia	Wild	DQ335856[b]	Nolan and Cribb (2006a)
Phthinomita poulini	*Parupeneus barberinus*	A	SW Pacific, off Lizard Island, Australia	Wild	DQ335857[b]	Nolan and Cribb (2006a)
	Parupeneus bifasciatus	A	SW Pacific, off Lizard Island, Australia	Wild	DQ335858[b]	Nolan and Cribb (2006a)
	Parupeneus cyclostomus	A	SW Pacific, off Lizard Island, Australia	Wild	DQ335859[b]	Nolan and Cribb (2006a)
Phthinomita robertsthomsoni	*Siganus argenteus*	A	SW Pacific, off Lizard Island, Australia	Wild	DQ335860[b]	Nolan and Cribb (2006a)
Phthinomita sasali	*Siganus doliatus*	A	Indo-West Pacific, Palau	Wild	DQ335861[b]	Nolan and Cribb (2006a)
Phthinomita symplocos	*Siganus lineatus*	A	SW Pacific, off Lizard Island, Australia	Wild	DQ335867[b]	Nolan and Cribb (2006a)

Continued

Table 1.1 GenBank DNA sequences for fish blood flukes—cont'd

Parasite	Host	Stage	Locality	Setting	GenBank accession numbers				Reference(s)
					18S	28S	ITS2	COI	
Plethorchis acanthus	*Mugil cephalus*	A	Brisbane River, Australia	Wild	AY222096[a]	AY222178[c]			Olson et al. (2003)
		A	SW Pacific, off Heron Island, Australia	Wild			AY465875[b]		Nolan and Cribb (2006a)
Psettarium (as *Paradeontacylix*) *sinensis*	*Takifugu rubripes*	A	Fuzhou, China	Cultured	EU081899[a]		EU082007[b]		Chen et al. (2008)
		A	ns	ns		EU368853[c]			Unpublished[d]
Sasala nolani	*Arothron meleagris*	A	S Pacific, off Moorea, French Polynesia	Wild	AY157184[a]	AY157174[a]			Lockyer et al. (2003a)
Skoulekia meningialis	*Diplodus vulgaris*	A	Mediterranean Sea, off Valencia, Spain	Wild	FN652294[a]	FN652293[c]	FN652292[b]		Alama-Bermejo et al. (2011)

[a]Near-complete nucleotide sequence.
[b]Complete nucleotide sequence.
[c]Partial nucleotide sequence.
[d]C.M. Chen, Y.Y. Wang, J.X. Wen (Submitted to GenBank in 04 August 2007, unpublished data).
ns, not specified; A, adult; C, cercaria; E, egg.

Table 1.2 Molecular studies of fish blood flukes

References	Approach	Taxonomic level	18S	ITS2	28S	COI	Sets of PCR primers[a]	
							PCR + sequencing	Additional sequencing
Barker and Blair (1996)	DP	Interfamily	✓		✓		3, 4, 59, 60	133–135
Snyder and Loker (2000)	DP	Interfamily			✓		57, 58	137–139
Cribb et al. (2001)	DP	Interfamily	✓				1, 2, 5–8	None
Littlewood and Olson (2001)	DP	Interfamily	✓				6, 7, 9–43	None
Olson et al. (2003)	DP	Interfamily	✓		✓		10, 42, 57, 61	146, 148, 149[b], 150
Lockyer et al. (2003a)	DP	Interfamily	✓		✓		1, 2, 63–68	146, 147, 149–163
Lockyer et al. (2003b)	DP	Interfamily	✓		✓	✓	1, 2, 63–68, 121, 122	146, 147, 149–169
Nolan and Cribb (2004a)	LC/SD	Intrafamily		✓			93[b]–95	None
Nolan and Cribb (2004b)	SD	Intrafamily		✓			93[b]–95	None
Snyder (2004)	DP	Interfamily	✓		✓		9, 42, 63, 64	140–147, 149[b], 150
Brant et al. (2006)	LC/SD	Intrafamily	✓		✓	✓	9, 42, 44–54, 58, 63, 64, 69–92, 123–130	140–147, 149[b], 150
Nolan and Cribb (2006a)	SD	Intrafamily		✓			94, 95	None

Continued

Table 1.2 Molecular studies of fish blood flukes—cont'd

References	Approach	Taxonomic level	18S	ITS2	28S	COI	Sets of PCR primers: PCR + sequencing	Sets of PCR primers: Additional sequencing
Nolan and Cribb (2006b)	SD	Intrafamily		✓			94, 95	None
Aiken et al. (2007)	SD	Intrafamily		✓	✓		57, 62, 93[b]–95	None
Ogawa et al. (2007)	SD	Intrafamily		✓			94, 95	None
Bullard et al. (2008)	DP	Intrafamily	✓				9, 42	140–143
Chen et al. (2008)	SD	Intrafamily	✓	✓			55, 56, 94, 95	None
Holzer et al. (2008)	SD	Intrafamily		✓	✓		63, 64, 93[b], 94	None
Repullés-Albelda et al. (2008)	SD	Intrafamily		✓	✓	✓	63, 64, 93[b], 94, 131, 132	152
Cribb et al. (2011)	DP/LC	Intrafamily		✓	✓		57, 62[b], 94, 95	None
Ogawa et al. (2011)	SD	Intrafamily		✓	✓		63, 64, 93[b], 94	None
Alama-Bermejo et al. (2011)	SD	Intrafamily	✓	✓	✓		10, 42, 63, 64, 94, 95	136, 152
Bray et al. (2012)	DP/SD	Intrafamily	✓		✓		1, 2, 63–68	146, 147, 149–163
Hernández-Orts et al. (2012)	SD	Intrafamily	✓		✓		10, 42, 63, 64	136, 152
Kirchhoff et al. (2012)[d]	SD	Intrafamily		✓				

Norte dos Santos et al. (2012)[d]	SD	Intrafamily	✓			
Shirakashi et al. (2012)	SD	Intrafamily	✓		96–101	None
Shirakashi et al. (2013)	SD	Intrafamily	✓	✓	63, 64, 93[b], 94	152
Yong et al. (2013)	SD	Intrafamily	✓		94, 95	None
Polinski et al. (2013)	SD	Intrafamily	✓		102–120[c]	None
Polinski et al. (2014)	SD	Intrafamily	✓		108, 109, 113–115, 120[c]	None
Sugihara et al. (2014)	LC/SD	Intrafamily	✓	✓	93[b], 94, 63, 64	

[a]See Table 1.3 for details about primer IDs in this chapter.
[b]Primer modified from original reference (see Table 1.3).
[c]Primers and probes for real-time qPCR detection and restriction-free cloning.
[d]Authors did not report primers used.
Deep phylogeny, DP; life cycles, LC; species differentiation, SD.

Table 1.3 Oligonucleotide primers for fish blood flukes (Digenea: Aporocotylidae)

Primer ID (direction)			
Original	**In article**	**Sequence 5′ to 3′**	**References**
Amplification and sequencing			
18S rDNA			
A (F)	1	AMCTGGTGGATCCTGCCG	Medlin et al. (1988)
B (R)	2	TGATCCATCTGCAGGTTCACCT	Medlin et al. (1988)
SB8	3	GGGTGGA TTTATTAGAACAG	Barker and Blair (1996)
PB	4	CCGTCAATTCMTTTRAGTTT	Barker and Blair (1996)
400F (F)	5	TCCGGAGAGGGAGCCTGA	Littlewood et al. (2000)
600R (R)	6	ACCGCGGCKGCTGGCACC	Littlewood et al. (2000)
1270F (F)	7	ACTTAAAGGAATTGACGG	Littlewood et al. (2000)
1630R (R)	8	TAAGGGCATCACAGACCTG	Littlewood et al. (2000)
18SE (alias 18S-A) (F)	9	CCGAATTCGTCGACA ACCTGGTTGATCCTGCCAGT	Littlewood and Olson (2001)
Worm A (F)	10	GCGAATGGCTCATTAAATCAG	Littlewood and Olson (2001)
18S-7 (F)	11	GCCCTATCAATTTGTTGGTA	Littlewood and Olson (2001)
18S-10 (R)	12	TACCATCGACAGTTGATAGGGC	Littlewood and Olson (2001)
300F (F)	13	AGGGTTCGATTCCGGAG	Littlewood and Olson (2001)
400R (alias 300R) (R)	14	TCAGGCTCCCTCTCCGGA	Littlewood and Olson (2001)

Cestode-1 (alias CEST1R) (R)[a]	15	TTTTTCGTCACTACCTCCCC	Littlewood and Olson (2001)
600F (F)	16	GGTGCCAGCMGCCGCGGT	Littlewood and Olson (2001)
18S-8 (F)	17	GCAGCCGCGGTAACTCCAGC	Littlewood and Olson (2001)
Pace-A (F)	18	GAGTTACCGCGGCTGCTG	Littlewood and Olson (2001)
18S-9 (F)	19	TTTGAGTGCTCAAAGCAG	Littlewood and Olson (2001)
930F (F)	20	GCATGGAATAATGAAATAGG	Littlewood and Olson (2001)
18S-A27 (R)	21	CCATACAAATGCCCCCGTCTG	Littlewood and Olson (2001)
Ael-5 (F)	22	TGTTTTCATTGATCAGGAGC	Littlewood and Olson (2001)
1100F (F)[b]	23	CAGAGTTTCGAAGACGATC	Littlewood and Olson (2001)
1100R (R)	24	GATCGTCTTCGAACCTCTG	Littlewood and Olson (2001)
Ael-3 (R)	25	GTATCTGATCGTCTTCGAAA	Littlewood and Olson (2001)
Pace-B (R)	26	CCGTCAATTCCTTTAAGTTT	Littlewood and Olson (2001)
1270R (R)	27	CCGTCAATTCCTTTAAGT	Littlewood and Olson (2001)
Pace-BF (F)	28	AAACTTAAAGGAATTGACGG	Littlewood and Olson (2001)
18S-11F (F)	29	AACGGCCATGCACCACCACCC	Littlewood and Olson (2001)
1262R (alias 1055R) (R)	30	CGGCCATGCACCACC	Littlewood and Olson (2001)
18S-11F (F)	31	GGGTGGTGGTGCATGGCCGTT	Littlewood and Olson (2001)
1262F (alias 1055F) (F)	32	GGTGGTGCATGGCCG	Littlewood and Olson (2001)
18S-2 (F)	33	ATAACAGGTCTGTGATGCCCTTAGA	Littlewood and Olson (2001)

Continued

Table 1.3 Oligonucleotide primers for fish blood flukes (Digenea: Aporocotylidae)—cont'd

Primer ID (direction)			
Original	**In article**	**Sequence 5′ to 3′**	**References**
1200F (F)	34	CAGGTCTGTGATGCCC	Littlewood and Olson (2001)
18S-3 (R)	35	TCTAAGGGCATCACAGACCTGTTAT	Littlewood and Olson (2001)
1200R (R)	36	GGGCATCACAGACCTG	Littlewood and Olson (2001)
18S-5 (F)	37	CCCTTTGTACACACCG CCCGTCGCT	Littlewood and Olson (2001)
1400F (F)	38	TGYACACACCGCCCGTC	Littlewood and Olson (2001)
18S-4 (R)	39	AGCGACGGGCGGTGTGTAC	Littlewood and Olson (2001)
1400R (R)	40	ACGGGCGGTGTGTAC	Littlewood and Olson (2001)
Cestode-6 (R)	41	ACGGAAACCTTGTTACGACT	Littlewood and Olson (2001)
Worm B (R)	42	CTTGTTACGACTTTTACTTCC	Littlewood and Olson (2001)
18S-F (alias 18S-B) (F)	43	CCAGCTTGATCCTTCTGCA GGTTCACCTAC	Littlewood and Olson (2001)
JB1 (F)	44	CCAACCTGGTTGATCCTGCCAGT	Morgan et al. (2003)
18SA (F)	45	AACCTGGTTGATCCTGCCAGT	Morgan et al. (2003)
18SF2.1 (F)	46	ATCTAAGGAAGGCAGCAGGCG	Morgan et al. (2003)
18SF1 (F)	47	CGGGACTCAATTGAGGCTCCGT	Morgan et al. (2003)
18SF2 (F)	48	ACTTTGAACAAATTTGAGTGCTCA	Morgan et al. (2003)
18H (F)	49	GCTGAAACTTAAAGGAATTGA	Morgan et al. (2003)

18SR0 (R)	50	CGCGGCTGCTGGCACCAGACTTGCC	Morgan et al. (2003)
18SR1(R)	51	CAGTGTCCGGGCCGGGTGAG	Morgan et al. (2003)
18J (R)	52	GGGCATCACAGACCTGTTATTG	Morgan et al. (2003)
R18A (R)	53	GATCCTTCCGCAGGTTCACCTACG	Morgan et al. (2003)
18SB (R)	54	TGATCCTTCTGCAGGTTCACCTAC	Morgan et al. (2003)
F2 (F)	55	GCCATGCATGTCCAAGTACATAC	Chen et al. (2008)
R2 (R)	56	TCGCTAAACCATTCAATCGGTAG	Chen et al. (2008)
28S rDNA			
LSU5 (F)	57	TAGGTCGACCCGCTGAAYTTAAGCA	Littlewood (1994)
LSU3 (R)	58	TAGAAGCTTCCTGAGGGAAACTTCGG	Littlewood (1994)
Ns[c]	59	GATTACCCGCTGAACTTAAGCATAT	Barker and Blair (1996)
Ns[a]	60	GCTGCATTCACAAACACCCCGACTC	Barker and Blair (1996)
1500R (R)	61	GCTATCCTGAGGGAAACTTCG	Olson et al. (2003)
EC-D2 (alias ECD2, ECD-2) (R)[bd]	62	CCTTGGTCCGTGTTTCAAGACGGG	Littlewood et al. (1997)
U178 (F)	63	GCACCCGCTGAAYTTAAG	Lockyer et al. (2003a)
L1642 (R)	64	CCAGCGCCATCCATTTTCA	Lockyer et al. (2003a)
U1148 (F)	65	GACCCGAAAGATGGTGAA	Lockyer et al. (2003a)
L2450 (R)	66	GCTTTGTTTTAATTAGACAGTCGGA	Lockyer et al. (2003a)
U1846 (F)	67	AGGCCGAAGTGGAGAAGG	Lockyer et al. (2003a)

Continued

Table 1.3 Oligonucleotide primers for fish blood flukes (Digenea: Aporocotylidae)—cont'd

Primer ID (direction)			
Original	**In article**	**Sequence 5′ to 3′**	**References**
L3449 (R)	68	ATTCTGACTTAGAGGCGTTCA	Lockyer et al. (2003a)
28SF6 (F)	69	GCACCCGCTGAAYTTAAG	Morgan et al. (2003)
C1 (F)	70	ACCCGCTGAATTTAAGCAT	Morgan et al. (2003)
ITS2.2F (F)	71	GCGGAGGAAAAGAAACTAAC	Morgan et al. (2003)
28SF4 (F)	72	AGTACCGTGAGGGAAAGTTG	Morgan et al. (2003)
28SF3 (F)	73	CGAAACCCAAAGGCGCAGTGA	Morgan et al. (2003)
28SF7 (F)	74	CCCGAAAGATGGTGAACTATGCTT	Morgan et al. (2003)
28SF9 (F)	75	GTATAGGGGCGAAAGACTAATCG	Morgan et al. (2003)
28SF10 (F)	76	AGCAGGACGGTGGCCATGGAAG	Morgan et al. (2003)
28SF8 (F)	77	AGGCCGAGGTGGAGAAGGGTTC	Morgan et al. (2003)
28SF11 (F)	78	TACCCATATCCGCAGCAGGTCTC	Morgan et al. (2003)
28SF12 (F)	79	AAACGGCGGAGTAACTATGA	Morgan et al. (2003)
28SF13 (F)	80	ATGGATGTAGTATAGGTGGGAGC	Morgan et al. (2003)
28SF14 (F)	81	AAGAGGTGTCAGAAAAGTTACC	Morgan et al. (2003)
D2 (R)	82	TGGTCCGTGTTTCAAGAC	Morgan et al. (2003)
28SR3 (R)	83	CTCAGGCATAGTTCACCATC	Morgan et al. (2003)
28SR1 (R)	84	AGCGCCATCCATTTTCAGGG	Morgan et al. (2003)

28SR6 (R)	85	GACCAAGTGCAGCTTGCCCTC	Morgan et al. (2003)
28SR9 (R)	86	AGACCTGCTGCGGATATGGGT	Morgan et al. (2003)
28SR7 (R)	87	GCTTTGTTTTAATTAG ACAGTCGGATTC	Morgan et al. (2003)
28SR10 (R)	88	GGGAATCTCGTTAATCCATTCA	Morgan et al. (2003)
28SR11 (R)	89	TCACCATAGGACACCCGCGT	Morgan et al. (2003)
28SR12 (R)	90	TGAACCTGCGGTTCCTCTCGTA	Morgan et al. (2003)
28SR13 (R)	91	ACTTAGAGGCGTTCAGTCTTAA	Morgan et al. (2003)
28SR8 (R)	92	ATTCTGACTTAGAGGCGTTCA	Morgan et al. (2003)
ITS2 rDNA			
3S (F)[e]	93	GGTACCGGTGGATCAC TCGGCTCGTG	Bowles et al. (1993)
ITS2.2 (R)	94	CCTGGTTAGTTTCTTT TCCTCCGC	Cribb et al. (1998)
GA1 (F)	95	AGAACATCGACATCTTGAAC	Anderson and Barker (1998)
CO-F (F)	96	GCTATTCCTAGATGTTTACG	Shirakashi et al. (2012)
CO-R (R)	97	GCAAAGAAACATTGCATCG	Shirakashi et al. (2012)
CH-F (F)	98	TTTTCCTAAATGTGTGTGCA	Shirakashi et al. (2012)
CH-R (R)	99	AGGCAACAAGTATCAAAACA	Shirakashi et al. (2012)
BF-F (F)	100	GGAAATTGTGCYACCTGGCA	Shirakashi et al. (2012)

Continued

Table 1.3 Oligonucleotide primers for fish blood flukes (Digenea: Aporocotylidae)—cont'd

Primer ID (direction)			
Original	**In article**	**Sequence 5′ to 3′**	**References**
BF-R (R)	101	AGCACAAGCCGCTACCA	Shirakashi et al. (2012)
Cfor_F	102	TGATTGCTTGCTTTTTCTCGAT	Polinski et al. (2013)
LCfor_F	103	TGCACAATTCACGACT CACGATCCACACGGT CTCGCACTGGCACGGGTGA TTGCTTGCTTTTTCTCGAT	Polinski et al. (2013)
Cfor_R	104	TATCAAAACATCAATCGACATC	Polinski et al. (2013)
RF_Cfor_F[f]	105	CGACTCACTATAGG GCAGATCTTCGAATGATTGCTTGCTTTT TCTCGATATG	Polinski et al. (2013)
RF_Cfor_R[f]	106	GGCCTTGACT AGAGGGTACCAGATATC AAAACATCAATCG ACATCTCA	Polinski et al. (2013)
Cori_F	107	TGCTTGCTATTCCTAGATGTTTAC	Polinski et al. (2013)
L_Cori_F	108	TGCACAATT CACGACTCACGATCATCC GCTCCGACGACACGAAC GGGTGCTTGCTATTCCT AGATGTTTAC	Polinski et al. (2013)
Cori_R	109	AACAACTATACTAAGCCACAA	Polinski et al. (2013)

RF_Cori_F[f]	110	CGACTCACTATAGGGC AGATCTTCGAATG CTTGCTATTCCT AGATGTTTACG	Polinski et al. (2013)
RF_Cori_R[f]	111	GGCCTTGACTAG AGGGTACCAGAAAC AACTATACTAAGCCACAACCT	Polinski et al. (2013)
Copt_F	112	TTCCTAAATGTGTGTGCA	Polinski et al. (2013)
L_Copt_F	113	TGCACAATTCACGA CTCACGATCATCC GCTCCGACGACACGAACGGGTT CCTAAATGTGTGTGCA	Polinski et al. (2013)
Copt_R	114	TCAAAACATCAATCGACACT	Polinski et al. (2013)
RF_Copt_F[f]	115	CGACTCACTATAGGGC AGATCTTCGAATTCC TAAATGTGTGTG CATTTGTG	Polinski et al. (2013)
RF_Copt_R[f]	116	GGCCTTGACTAG AGGGTACCAGATCAAAA CATCAATCGACA CTTCAC	Polinski et al. (2013)
L_UP	117	GCACAATTCACGACTCACGA	Polinski et al. (2013)
L_FAM_1[g]	118	FAM-CCACACGGTCTC GCACTGGC-BHQ1	Polinski et al. (2013)

Continued

Table 1.3 Oligonucleotide primers for fish blood flukes (Digenea: Aporocotylidae)—cont'd

Primer ID (direction)			
Original	**In article**	**Sequence 5′ to 3′**	**References**
L_HEX_1[g]	119	HEX-CCACACGGTCTCG CACTGGC-BHQ1	Polinski et al. (2013)
L_HEX_2[g]	120	HEX-CATCCGCTCCGAC GACACGA-BHQ1	Polinski et al. (2013)
COI mtDNA			
Cox1_schist_5k (F)	121	TCTTTRGATCATAAGCG	Lockyer et al. (2003b)
Cox1_schist_3k (R)	122	TAATGCATMGGA AAAAAACA	Lockyer et al. (2003b)
CO1F5 (F)	123	TTGRTTTGTYTCTTTRGATC	Morgan et al. (2003)
CO1F6 (F)	124	TTTGTYTCTTTRGATCATAAGCG	Morgan et al. (2003)
CO1F4 (F)	125	ATTTGGWACTGCTTTTTTTGAGCC	Morgan et al. (2003)
CO1F3 (F)	126	CATTTATTTTGGTTTTTTGGTCA	Morgan et al. (2003)
CO1R9 (R)	127	TTDTTHCTTADABTCATACA	Morgan et al. (2003)
CO1R8 (R)	128	CCAAYCATRAACATATGATG	Morgan et al. (2003)
CO1R4 (R)	129	ACCTAAATAATGCATAGGAAA	Morgan et al. (2003)
CO1R5 (R)	130	GATCATARCAWCTWACACGACG	Morgan et al. (2003)
JB3 (F)	131	TTTTTTGGGCATCCTGAGGTTTAT	Bowles et al. (1993)
JB4.5 (R)	132	TAAAGAAAGAACATAATGAAAATG	Bowles et al. (1993)

Additional sequencing[h]			
18S rDNA			
SB9	133	TTTCACCTCTAACACCGC	Barker and Blair (1996)
SB3	134	GGAGGGCAAGUCUGGUGC	Barker and Blair 1996)
A27	135	CCATACAAATGCCCCCGTCTG	Barker and Blair (1996)
Lin 3 (F)	136	GCGGTAATTCCAGCTCCA	Lin et al. (1999)
D2F (F)	137	CTTTGAAGAGAGAGTTC	Littlewood (1994)
D3RM (R)	138	GCATAGTTCACCATCTTTC	Littlewood and Johnston (1995)
D4AR (R)	139	CCGTGTTTCAAGACGGG	Littlewood and Johnston (1995)
388F (F)	140	AGG GTT CGA TTC CGG AG	Littlewood and Olson (2001)
1100F (F)[b]	141	CAGAGTTTCGAAGACGATC	Littlewood and Olson (2001)
CEST1R (R)[b]	142	TTTTTCGTCACTACCTCCCC	Littlewood and Olson (2001)
1270R (R)	143	CCGTCAATTCCTTTAAGT	Littlewood and Olson (2001)
28S rDNA			
Dig12 (F)	144	AAGCATATCACTAAGCGG	Tkach et al. (1999)
LSU1500R (R)	145	GCTATCCTGAGGGAAACTTCG	Tkach et al. (1999)
300F (F)	146	CAAGTACCGTGAGGGAAAGTTG	Lockyer et al. (2003a)
300R (R)	147	CAACTTTCCCTCACGGTACTTG	Lockyer et al. (2003a)
400R (R)	148	GCA GCT TGA CTA CAC CCG	Olson et al. (2003)

Continued

Table 1.3 Oligonucleotide primers for fish blood flukes (Digenea: Aporocotylidae)—cont'd

Primer ID (direction)			
Original	**In article**	**Sequence 5′ to 3′**	**References**
EC-D2 (alias ECD2, ECD-2) (R)[b]	149	CCTTGGTCCGTGTTTCAAGACGGG	Littlewood et al. (1997)
900F (F)	150	CCGTCTTGAAACACGGACCAAG	Lockyer et al. (2003a)
1200F (F)	151	CCCGAAAGATGGTGAACTATGC	Lockyer et al. (2003a)
1200R (alias LSU1200R) (R)	152	GCATAGTTCACCATCTTTCGG	Lockyer et al. (2003a)
1600F (F)	153	AGCAGGACGGTG GCCATGGAAG	Lockyer et al. (2003a)
U2229 (F)	154	TACCCATATCCGC AGCAGGTCT	Lockyer et al. (2003a)
L2230 (R)	155	AGACCTGCTGCG GATATGGGT	Lockyer et al. (2003a)
U2562 (F)	156	AAACGGCGGGAGTAACTATGA	Lockyer et al. (2003a)
L2630 (R)	157	GGGAATCTCGTTAATCCATTCA	Lockyer et al. (2003a)
U2771 (F)	158	AGAGGTGTAGGATARGTGGGA	Lockyer et al. (2003a)
L2984 (R)	159	CTGAGCTCGCCTTAGGACACCT	Lockyer et al. (2003a)
U3119 (F)	160	TTAAGCAAGAGGTGTCAGAAAAGT	Lockyer et al. (2003a)
U3139 (F)	161	AAGTTACCACAGGGATAACTGGCT	Lockyer et al. (2003a)
LSU3_4160 (R)	162	GGTCTAAACCCAGCTCACGTTCCC	Lockyer et al. (2003a)
L3358 (R)	163	AACCTGCGGTTCCTCTCGTACT	Lockyer et al. (2003a)

COI mtDNA			
CO1560Fa (F)	164	TTTGATCGTAAATTTGGTAC	Lockyer et al. (2003b)
CO1560Fb (F)	165	TTTGATCGGAATTTTGGTAC	Lockyer et al. (2003b)
CO1560R (R)	166	GCAGTACCAAATTTACGATC	Lockyer et al. (2003b)
CO1800F (F)	167	CATCATATGTTTATGGTTGG	Lockyer et al. (2003b)
CO1800Ra (R)	168	CCAACCATAAACATATGATG	Lockyer et al. (2003b)
CO1800Rb (R)	169	CCAACCATAAACATGTGATG	Lockyer et al. (2003b)

[a]Primer only identified by its 3′ starting position in *S. mansoni* sequence (i.e., position 278) (see Barker and Blair, 1996).
[b]Primers presented in "amplification and sequencing" and "additional sequencing" category in order to better characterize their distinctive role in different articles (see Table 1.2).
[c]Primer only identified by its 3′ starting position in *S. mansoni* sequence (i.e. position 45) (see Barker and Blair, 1996).
[d]Snyder (2004), Cribb et al. (2011) employed an oligonucleotide sequence modified from original reference with the subtraction of a cytosine in 5′ end.
[e]Nolan and Cribb (2004a), Nolan and Cribb (2004b), Holzer et al. (2008), Ogawa et al. (2011), and Shirakashi et al. (2013) used an oligonucleotide sequence modified from original reference (i.e. 5′-GGTACCGGTGGATCAC**GT**GGCT**A**GTG-3′).
[f]Restriction-free bridging primers.
[g]Real-time PCR probe.
[h]Primers were classified as "additional sequencing" just when this was specified in the original reference.
ns, not specified.

Bray et al., 2012; Bullard et al., 2008; Cribb et al., 2011; Holzer et al., 2008; Nolan and Cribb, 2004a, 2006a,b; Ogawa et al., 2011).

FBFs are of rapidly emerging interest to ecology and evolutionary biology because some lineages may have coevolved with the major lineages of nontetrapod craniates (Bullard et al., 2008). They are relevant to medical researchers because they are the likely immediate ancestor to the tetrapod blood flukes (Brant et al., 2006), including those that cause schistosomiasis (Schistosomatidae) (op. cit.). They are of critical concern to aquatic animal health personnel who regard them as serious pathogens of high-value fishes kept in marine (Cribb et al., 2011; Ogawa et al., 2007) and freshwater (Kirk, 2012; Meade, 1967; Meade and Pratt, 1965; Schell, 1974) aquaculture operations (Bullard and Overstreet, 2002, 2008). Taxonomists also are interested in them because the rate of new species discovery is proportionally high relative to that of other fish trematode groups (Cribb and Bray, 2011). Yet, FBFs remain among the most poorly understood of trematode families, and ambiguity regarding their interrelationships obstructs a nuanced understanding of the evolutionary origins of flatworm parasitism in the blood of craniates, including the origin of schistosomes.

An abundance of recent molecular phylogenetic data supports the notion that FBFs are the direct aquatic counterparts and likely ancestors of the tetrapod blood flukes, turtle blood flukes "tetrapod blood flukes" = "turtle blood flukes + schistosomes" schistosomes. Yet they are one of the most poorly understood of trematode families (Cribb and Bray, 2011). They require a single intermediate host, a gastropod, bivalve, or polychaete, and mature in the blood and body cavity, rarely other sites (Alama-Bermejo et al., 2011), of fishes ranging worldwide in rivers (Bullard et al., 2008; Truong and Bullard, 2013), estuaries (Bullard, 2013), coral reefs (Nolan and Cribb, 2006a,b), and offshore epipelagic waters (Ogawa et al., 2010; Orélis-Ribeiro et al., 2013). FBFs presently comprise 136 accepted species assigned to 35 genera, 20 of which are monotypic (Fig. 1.1). They thus outnumber both schistosomes (94 species in 15 genera)[1] and spirorchiids (85 species in 19 genera).[2] Nearly half of the recognized FBF

[1] Basch (1991) counted 86 species and 13 genera; Khalil (2002) accepted *Jilinobilharzia* as valid, which includes two species; Müller and Kimmig (1994), Horák et al. (1998), and Kolářová et al. (2013) described three new species of *Trichobilharzia*; Attwood et al. (2002) and Hanelt et al. (2009) described two new species of *Schistosoma*; Aldhoun and Littlewood (2012) considered *Orientobilharzia*, a junior subjective synonym of *Schistosoma*; Kolářová et al. (2006) and Brant et al. (2013) proposed two new genera and species (*Allobilharzia visceralis* and *Anserobilharzia brantae* (syn. *Trichobilharzia brantae* Farr and Blankemeyer, 1956), respectively).

[2] Smith (1997b) listed 82 species of 21 genera; Platt (2002) accepted 19 genera; Tkach et al. (2009) and Platt and Sharma (2012) described 1 and 2 new species, respectively.

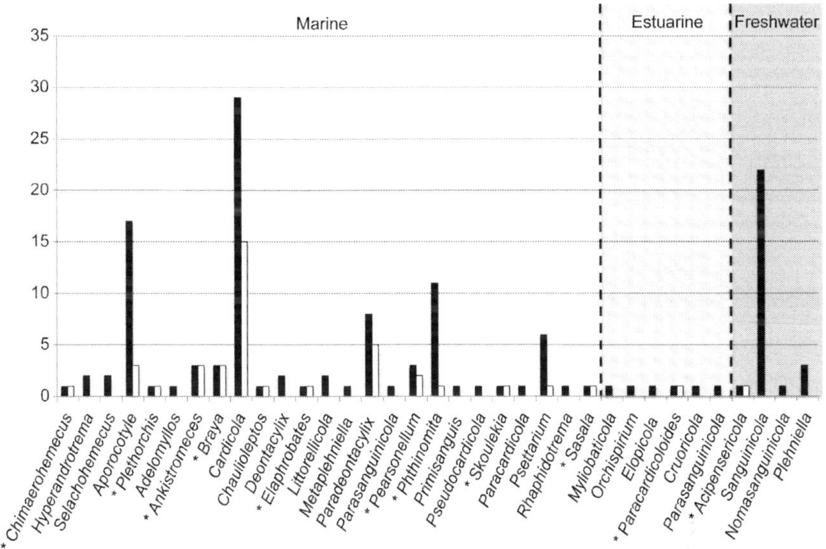

Figure 1.1 Number of accepted fish blood fluke species per accepted genus (black bars) and that has been characterized with nucleotide data (white bars). Categorized by general ecological providence (marine, estuarine, and freshwater) and presented from left to right in approximate phylogenetic order (basal to derived; based on Nelson, 2006) of the definitive host group infected by the species of that genus. *A genus whose type species has been sequenced.

genera (15 of 35) have been proposed since 2002 (Cribb and Bray, 2011; Smith, 2002), and many species have been described since 2002. Most recent descriptions have been conducted by workers in the western Pacific Ocean off Australia (T. Cribb et al.) and Japan (K. Ogawa et al.), the Mediterranean Sea (J.A. Raga et al.), and the Gulf of Mexico, Caribbean Sea, and northwestern Atlantic Ocean (S.A. Bullard et al.). The large number of monotypic genera (Fig. 1.1), the regional nature of described species, i.e., most species typically are known from single collections from a single geographic locality, and the large proportion of fishes that have yet to be critically examined for infections together indicate that many species of FBFs remain undiscovered in each of these regions and in adjacent waters.

FBFs are relevant to both basic research and applied research especially considering (i) host–parasite coevolution and (ii) health implications for aquacultured fishes. Regarding coevolution, FBFs exploit the spectrum of vertebrate lineages (Amemiya et al., 2013; Nelson, 2006), from the most primitive jawed (gnathostome) craniates (Chondrichthyes: sharks, rays, and chimaeras) (Bazikalova, 1932; Bullard and Jensen, 2008; Bullard et al., 2006; Madhavi and Rao, 1970; Orélis-Ribeiro et al., 2013; Short, 1954; van der Land, 1967) to the most highly derived of bony fishes (Pleuronectiformes and Tetraodontiformes) (Goto and Ozaki, 1929; Manter, 1940; Martin, 1960; Nolan and Cribb, 2004c; Ogawa et al., 2007; Yong and Cribb, 2011) (Figs. 1.1–1.3). However, phylogenetic studies of FBFs have not kept pace with discoveries of new species and proposals of new genera. Phylogenetic inferences including each accepted genus and relevant out-group exist for Schistosomatidae and paraphyletic "Spirorchiidae" (op. cit.), but no comparable phylogeny exists for FBFs. As such, the monophyly of the FBFs remains untested and, we argue, likely based on the assumption that all nontetrapod blood flukes are "fish blood flukes" and therefore members of a monophyletic Aporocotylidae. Hence, our understanding of the evolutionary origins of the blood flukes infecting terrestrial craniates remains somewhat opaque since we have not yet scrutinized deep phylogenetic interrelationships of FBFs. Available evidence from morphology (Bullard, in press; Bullard et al., 2006; Orélis-Ribeiro et al., 2013; Truong and Bullard, 2013) and molecular biology (Bullard et al., 2008; Cribb et al., 2011; present study) hints that blood flukes have coevolved with the major lineages of craniates, perhaps accompanying craniates onto land. Noteworthy is that no record of a blood fluke exists from any nontetrapod lineage of Sarcopterygii, i.e., coelacanths and lungfishes (Fig. 1.2E), which represents an obvious gap in our knowledge regarding the natural distribution and evolutionary expansion of these blood parasites in aquatic craniates (mostly fishes) and terrestrial craniates (tetrapods) (Amemiya et al., 2013; Nelson, 2006; Fig. 1.2E). With that suspected, long-shared evolutionary history between these flukes and their fish hosts as a backdrop, we argue that a better understanding of the evolutionary interrelationships and natural history of FBFs will underpin new approaches for further understanding those attributes in their putative descendants, the tetrapod blood flukes.

The study of FBF infections has applications to aquaculture because FBFs debilitate or kill fish within intensive culture systems (Bullard and Overstreet, 2002, 2008; Hardy-Smith et al., 2012; Ishimaru et al., 2013). Requirements to survey nearby invertebrates and fishes for infections and manage diseases in those settings, especially in offshore net pens ("sea cages"), have hastened FBF life cycle and epidemiological studies (Alama-Bermejo et al., 2011; Cribb et al., 2011; Hayward et al., 2010; Holzer et al., 2008; Ogawa et al., 2007, 2011; Shirakashi et al., 2012). In fact,

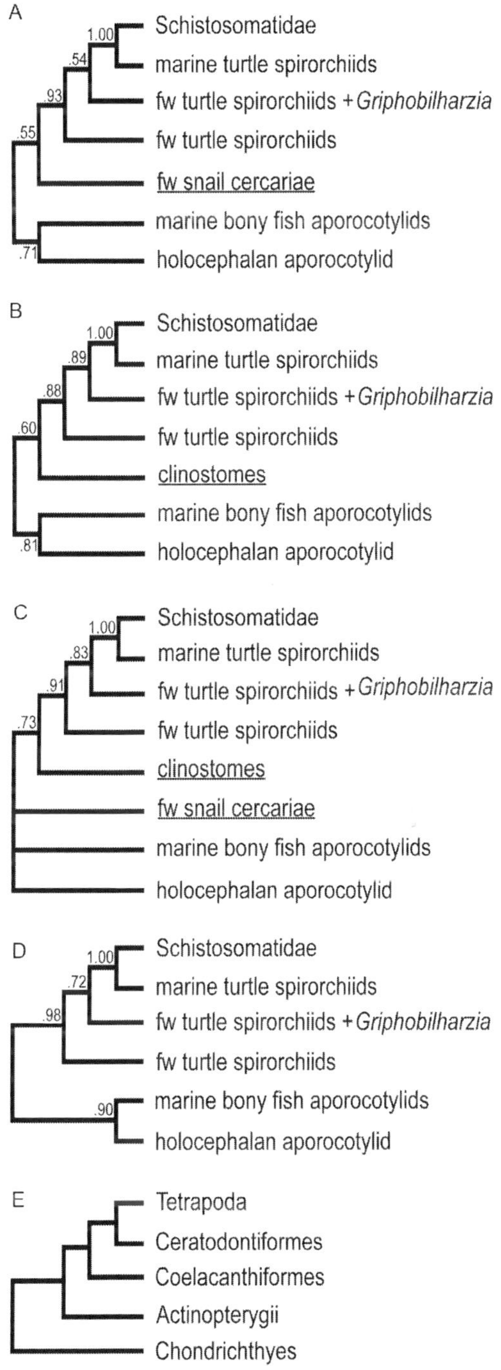

Figure 1.2 Simplified phylogenetic relationships of blood flukes (Bayesian inference, partial 28S sequences (see text for out-groups), and posterior probability aside

the intensity of research activities focused on fish pathogenic FBFs in aquaculture has greatly increased our knowledge of their biology and life cycles. The tuna blood fluke *Cardicola forsteri*, which matures in commercially prized southern bluefin tuna (*Thunnus maccoyii*), has arguably become the most intensively studied of marine fish trematodes. Moreover, as aquaculture technologies expand to include more fish species cultured under a wider diversity of freshwater, marine, and estuarine environments, novel FBF pathogens may emerge that constrain those sectors of the aquaculture industry.

We review published molecular biological studies of FBFs in the areas of life history, taxonomy, and phylogenetics. We also provide a new phylogenetic analysis for the flatworms infecting the blood vascular system of craniates and discuss the relevance of FBFs to studies of tetrapod blood flukes (spirorchiids and Schistosomatidae).

2. LIFE HISTORY

FBF cercariae are minute, are difficult to identify morphologically, and typically have an extremely low prevalence of infection among invertebrate intermediate host populations (Cribb et al., 2011). Perhaps as a result of these attributes, the intermediate host(s) is unknown for all but a few FBF species: several species of the freshwater genus *Sanguinicola* (see Bullard and Overstreet, 2008) and the marine species *Aporocotyle simplex* (see Køie, 1982; Køie and Petersen, 1988, experimental infections; no molecular markers applied), *Paracardicoloides yamagutii* (see Nolan and Cribb, 2004a; see succeeding text), *C. forsteri* (see Cribb et al., 2011; see succeeding text), and *Cardicola opisthorchis* (see Sugihara et al., 2014; see succeeding text). ITS2 rDNA sequences have been used to match morphologically

Figure 1.2—Cont'd each node) that (A) include the freshwater gastropod (snail) cercariae (Brant et al., 2006; Olson et al., 2003) and exclude clinostomes, (B) include clinostomes and exclude the freshwater snail cercariae, (C) include both freshwater snail cercariae and clinostomes, and (D) exclude both freshwater snail cercariae and clinostomes. (E) Simplified phylogeny for Gnathostomata (craniates other than Myxiniformes (hagfishes) + Petromyzontiformes (lampreys)) based on Nelson (2006) and Amemiya et al. (2013), showing the position of sharks, skates, rays, and chimaeras (Chondrichthyes); ray-finned fishes (Actinopterygii); and the three major lineages of Sarcopterygii: coelacanths (Coelacanthiformes), lungfishes (Ceratodontiformes), and terrestrial craniates and their descendants (Tetrapoda).

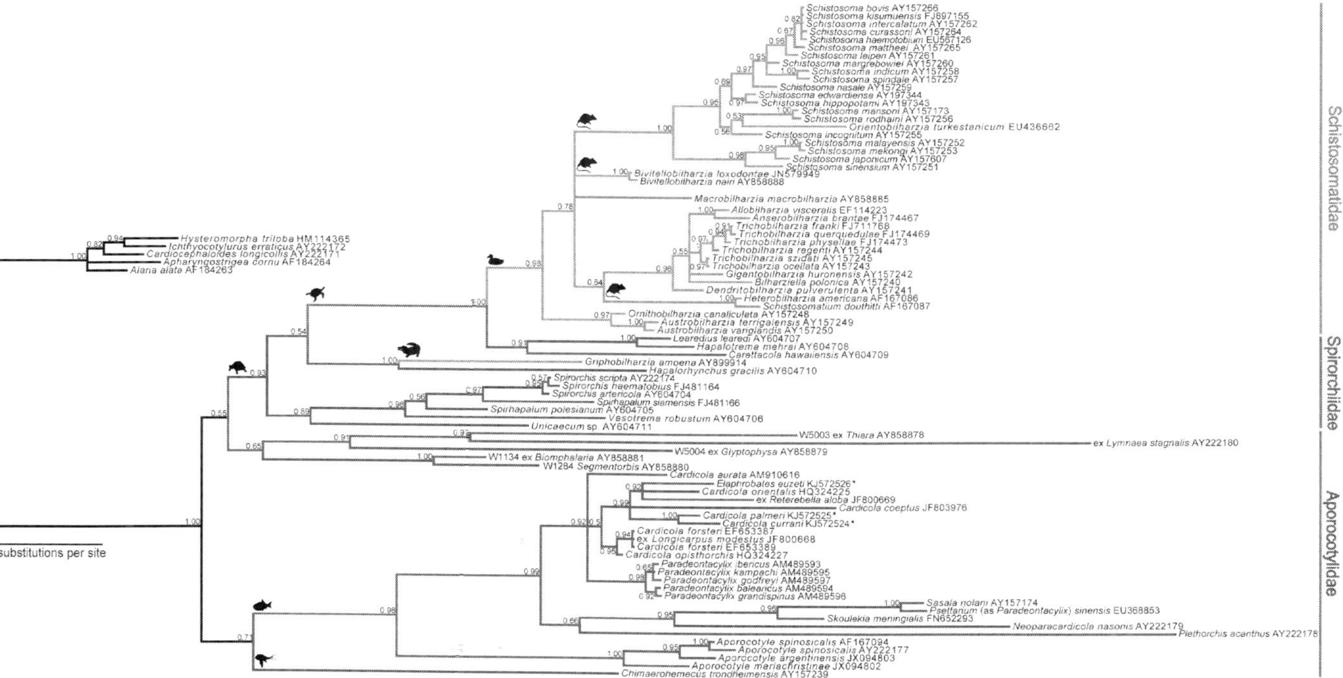

Figure 1.3 Phylogenetic relationships of blood flukes reconstructed by Bayesian inference and based on partial D1–D2 domains of 28S from 83 blood fluke taxa (majority rules consensus tree). Numbers aside tree nodes indicate posterior probability. *Taxa sequenced in the present study.

indeterminate larval FBF specimens, i.e., cercariae, sporocysts, and rediae, collected from invertebrate hosts with morphologically distinctive adult specimens from sympatric fish hosts. Expeditiously and inexpensively, molecular markers enable the rapid detection of infections; identification of the infective agent, by phylogenetic inference or by sequence homology to an already-sequenced taxon; and identification of intermediate hosts, thus linking life history stages of potentially pathogenic FBFs. Certainly, determining the identity of intermediate and definitive hosts comprises a critical first step in understanding the life cycle of FBFs. Beyond diagnostic approaches, however, much remains to be learned about specific details of the host–parasite relationship among FBFs and their polychaete, gastropod, and bivalve intermediate hosts. After all, their evolutionary history may well be influenced as much, or more so, by these intermediate hosts as their definitive hosts. As such, we argue that molecular studies should not wholly supplant classical experimental studies that make direct microscopy observations of larval and adult FBFs in host tissues subsequent to exposures of naive invertebrate and definitive fish hosts (Køie, 1982).

Much of the information concerning FBF life cycles originates from aquatic animal health programmes linked to commercial aquaculture: 17 of 109 (16%) of the available FBF sequences in GenBank derive from infections in marine aquaculture (Table 1.1). FBFs comprise one of the few trematode groups whose members can harm the definitive fish host as adults that occlude blood vessels and cause asphyxia, as eggs that damage or obstruct gill epithelia and branchial vessels, and as miracidia that hatch from eggs embedded in gill epithelium and bore out of the fish (Bullard and Overstreet, 2002, 2008). As such, their life cycles are of concern to commercial aquaculture operations because FBFs can kill or debilitate fish and cause economic losses in freshwater ponds, raceways, and offshore marine cages. Sequences sourced from nonadult FBFs infecting gastropod, bivalve, and polychaete hosts, which harbour FBF asexual reproductive stages—sporocyst, rediae, and cercariae—are far less numerous (4 of 109) (Table 1.1) than those from adult specimens infecting fish hosts.

In the first published FBF life cycle study determined by the application of a molecular method, Nolan and Cribb (2004a) documented two cercarial morphotypes ("type A" and "type B") infecting 80 of 11,314 (0.7%) specimens of the hydrobiid gastropod *Posticobia brazieri* in the tidal creeks of Queensland, Australia. ITS2 rDNA sequences from cercaria type A aligned (100% agreement) with adult specimens of *P. yamagutii* that were collected from the blood (the dorsal aorta, atrium, ventricle, gills, kidney, and blood vessels of the intestine and swim bladder) of speckled longfin eels *Anguilla reinhardtii*. In another study, responding to concerns about diseases associated with infections of *C. forsteri* infecting cultured southern bluefin tuna *T. maccoyii* off South Australia, Cribb et al. (2011) screened 9351 individuals of 11 bivalve, 2 gastropod, and 24 polychaete families for FBF infections. ITS2 rDNA sequence data derived from cercaria that infected *Longicarpus modestus* (Polychaeta: Terebellidae) aligned with 100% agreement with adult specimens of *C. forsteri* from the heart of nearby southern bluefin tuna in a net pen at Port Lincoln, Australia. They also sequenced the 28S rDNA fragment (721 bp in D1–D2 region) from cercaria of *C. forsteri* that infected *L. modestus*. Following the approach of Cribb et al. (2011), Sugihara et al. (2014) focused on polychaetes while examining 744 invertebrates for FBF infections in a culture site of the Pacific bluefin tuna *Thunnus orientalis*, off southern Japan. Sporocysts and cercaria of FBFs were found in five individuals of a terebellid polychaete (*Terebella*) collected from within the shell of dead barnacles taken from the substratum and from ropes and floats below the sea cage. ITS2 and 28S sequences from sporocysts were identical (100%) to those of adults of *C. opisthorchis* from the heart of cultured Pacific bluefin tuna.

Shirakashi et al. (2012) studied concurrent infections of *Cardicola orientalis* and *C. opisthorchis* in Pacific bluefin tuna. They used ITS2 rDNA sequence data to differentiate the crescent-shaped eggs of *C. opisthorchis* in the afferent filament artery from ovoid eggs of *C. orientalis* infecting the gill lamellae. They stated that species-specific PCR primers applied to gill tissue samples could complement histopathology and help diagnose infections before eggs and adults were numerous enough to be readily observed with light microscopy. Yong et al. (2013) used complete ITS2 rDNA sequence data to identify the FBF eggs lodged in the gill of five species of butterflyfish (Perciformes: Chaetodontidae) from the Great Barrier Reef as a single species *Cardicola chaetodontis* (a single base pair difference in two samples). FBF eggs are not infrequently observed in the gill epithelium and branchial arterioles of fishes during routine fish necropsies (SAB, personal observations), but in many instances, corresponding adult specimens that infect the blood of the individual fish are not recovered. Molecular techniques that effectively extract and amplify DNA from FBF eggs, which are minute, for example, the eggs of *C. chaetodontis* are 40–60 μm in total length, promise to reveal the presence of unnamed FBF species and hitherto undocumented hosts if resulting sequences are placed in a phylogenetic context with sequences from named species. The same can be said for sequences derived from cercariae (see succeeding text). Using quantitative polymerase chain reaction (qPCR), Norte Do Santos et al. (2012) identified the eggs of *C. forsteri* aggregated in the gill of ranched southern bluefin tuna. Similarly, Kirchhoff et al. (2012) applied this method to diagnose eggs of *C. forsteri* infecting *T. maccoyii*. Polinski et al. (2013) took the approach a step further by developing a sensitive and accurate real-time PCR technique that can also be used for the identification of *C. forsteri*, *C. orientalis*, and *C. opisthorchis* in nonlethal samples. No cross-species or host genomic amplification was detected in either method tested, i.e., SYBR-based qPCR and a common reporter TaqMan assay; however, their combined application improved the reliability to differentiate species. These methods have confirmed concurrent infections of *C. forsteri* and *C. orientalis* in *T. maccoyii*, also indicating the higher prevalence and distribution of *C. orientalis* in the host.

Recently, the usefulness of both developed techniques was supported to assess the pathological consequences of FBF infections. Polinski et al. (2014) combined qPCR with host gene immune transcription to quantify the amounts of DNA from tissues infected by *C. opisthorchis* and *C. orientalis*. They also quantified the temporal host immune response to those infections in caged Pacific bluefin tuna. Previous data documented adults and eggs of *C. orientalis* in the gill (Ogawa et al., 2010; Shirakashi et al., 2012), whereas adults and eggs of *C. opisthorchis* infected the heart and afferent branchial arteries, respectively (Ogawa et al., 2011; Shirakashi et al., 2012). Polinski et al. (2014) results, however, revealed a correlation of IgM transcription to the high quantities of "*C. orientalis* only" infections in the gill tissues but not to the DNA of *C. opisthorchis*, suggesting that such an immune response in this organ might be triggered by the presence of adults rather than of eggs. Moreover, high levels of DNA of *C. orientalis* in the heart were attributed to the presence of juvenile flukes. Although such methods cannot identify FBF life history stages or whether or not the flukes are alive or dead, this quantitative approach enables conjecture about infection intensity and infection status, which is itself a promising tool for future epidemiological studies involving wild or cultured fishes.

Brant et al. (2006) used phylogenetic inference (28S, 18S, and COI) in treating several unidentified putative FBF cercariae isolated from gastropods (Planorbidae and Thiaridae) in Uganda, Kenya, and Australia. They morphologically characterized the cercariae with photomicrographs and based upon the presence/absence of eyespots, fin folds on the cercarial tail and body, and tail shape. Although a consistent and phylogenetically coherent, morphology-based definition (diagnosis) of FBF cercariae is lacking, these cercariae are typically distinctive: minute, forktailed, without a ventral sucker, with a distinctive penetrating organ that likely comprises a specialized anterior sucker (Truong and Bullard, 2013), a fin fold on the cercarial body, and a fin fold present or absent on the tail furcae (Cribb et al., 2011). One of these cercaria (W5004) was especially morphologically bizarre, i.e., described as apharyngeate and furcocercous but with "extravagant lateral tail membranes and a pointed body shape", and none of these cercariae had all of the typical features of FBF cercariae (see preceding text). However, the combined 18S and 28S analysis of these cercariae clustered them with other FBF sequences (we presume that "Sanguinicolid sp." is that of "*Sanguinicola* cf. *inermis*" (AY222180)). For that reason and because they were morphologically bizarre relative to known FBF cercariae, these authors posited that a much greater morphological diversity of FBF cercariae exists than has been recognized in the literature to date. The taxonomic identity and phylogenetic significance of these sequences are further discussed later in the text.

3. TAXONOMY

Most available FBF DNA sequence information is derived from adult FBFs infecting the heart of marine fishes (Table 1.1). As of 14 February 2014, GenBank contained 109 FBF nucleotide sequences derived from 22 published papers and one sequence from an unpublished work (Tables 1.1 and 1.2). In total, these sequences represent 48 of 136 (35%) nominal FBFs assigned to 15 of 35 (43%) accepted genera and infecting 42 bony fish species and one chondrichthyan species (Tables 1.1 and 1.2). For most genera, very few or no species have been sequenced, and adults of only one freshwater FBF have been sequenced to date (Fig. 1.1; Bullard et al., 2008). ITS2 rDNA comprises 65 of 109 (59%) GenBank nucleotide sequences for 41 of 48 (85%) FBF species, 28S comprises 28 (26%) sequences for 20 (42%) species, 18S comprises 10 (9%) sequences for 10 (21%) species, and COI comprises 6 (6%) sequences for 5 (10%) species. Noteworthy also is that 29 of 48 (61%) FBF species in GenBank are represented only by a single gene: ITS2 (27 species), 28S (1), or 18S (1). Sequence data from the combination of 28S and ITS2 genes exist for 12 species and those of 18S and 28S genes exist for 9 species. Few FBF species have been characterized by more than two genes: 18S, 28S, and COI sequence data are available for *Chimaerohemecus trondheimensis*, and those for 18S, 28S, and ITS2 exist for *P. yamagutii*, *Psettarium* (as *Paradeontacylix*) *sinensis*,

Plethorchis acanthus, and *Skoulekia meningialis*. All ITS2 nucleotide sequences are represented by complete fragments (ITS2: ~390 bp), whereas the other fragments are depicted mostly by near-complete (18S, ~1800 bp; 28S, ~3700 bp; and COI, ~1100 bp) and partial sequences (28S, ~1300 bp and COI, ~400 bp) (Table 1.1). In addition, several sequences representing innominate FBFs have resulted from parasitological surveys of wild (Alama-Bermejo et al., 2011; Hernández-Orts et al., 2012; Nolan and Cribb, 2004a, 2006a,b) and cultured fishes (Holzer et al., 2008; Ogawa et al., 2007, 2011; Repullés-Albelda et al., 2008; see succeeding text) (Table 1.1).

Nolan and Cribb (2006a) characterized a high level of genetic diversity, interpreted as FBF species richness (two new species of *Ankistromeces* and nine new species of *Phthinomita*), among morphologically similar FBF specimens infecting siganid, labrid, and mullid reef fishes off Australia and Palau. They sequenced the replicates of complete ITS2 rDNA from specimens of each new species from a total of 29 host/parasite/locality combinations. The study revealed 19 distinct genotypes (having 1–41 base differences), which defined those species along with morphological characters—together representing the first published evidence of cryptic speciation among FBFs that otherwise would have been undetected or underestimated if morphology alone had been considered. This level of ITS2 sequence conservation is interesting since variation in the ITS1 is detectable within a species or an individual among other Digenea (Nolan and Cribb, 2005) and other fish parasitic platyhelminths, for example, ectoparasitic capsaline monogenoids (Bullard et al., 2011).

In a case of concurrent infection, Shirakashi et al. (2013) sequenced partial 28S and complete ITS2 rDNA regions to complement the morphological identification of *C. orientalis* and *C. forsteri* infecting the gill and heart, respectively, of ranched southern bluefin tuna off South Australia. This was the first report of *C. orientalis* infecting *T. maccoyii*, not only expanding the known geographic distribution of this FBF but also raising new concerns to the aquaculture industry.

To our knowledge, only one study has applied molecular tools to test a biogeographical hypothesis with FBFs. Aiken et al. (2007) tested whether or not parasites of epipelagic fishes are as geographically widespread and genetically similar as their hosts. Complete ITS2 rDNA sequences of *C. forsteri* from wild and cultured southern bluefin tuna off the Great Australian Bight and those from cultured Atlantic bluefin tuna *Thunnus thynnus* in the Mediterranean Sea off Spain matched 100%. A similar match was demonstrated among the sequences of *Cardicola* sp. from cultured Pacific bluefin tuna off

Mexico and cultured Atlantic bluefin tuna from Spain. Partial 28S rDNA sequences of *C. forsteri* from tunas off Australia and Spain differed by 1 bp (Holzer et al., 2008), perhaps explained by highly variable regions of the 28S (Olson and Tkach, 2005). Beyond basic research on biogeography, molecular characterizations and delineations of specific blood fluke strains in a given geographic area are relevant to the aquaculture industry since some of those strains may have different levels of pathogenicity to different host populations. Given how fishes in the food aquaculture industry and in the ornamental pet trade are transported globally, such information could be also relevant to bio-security and conservation biology of endemic fish populations.

4. PHYLOGENY

A phylogenetic hypothesis based on morphological or molecular data and including species from the majority of accepted FBF genera has yet to be published. Early studies incorporating FBF sequence data used 18S rDNA to root phylogenies testing monophyly and interrelationships of Schistosomatidae (Lockyer et al., 2003b; Snyder and Loker, 2000), Digenea (Cribb et al., 2001; Olson et al., 2003), and Platyhelminthes (Littlewood and Olson, 2001; Littlewood et al., 1999; Lockyer et al., 2003a). Later, studies using FBF sequences have focused on interrelationships among genera and species within the family (including ribosomal 18S, ITS2, and 28S plus mitochondrial COI genes; Tables 1.1 and 1.2). Most of these studies rely on already deposited GenBank sequences and add one or a few novel sequences to generate a phylogeny. Nearly all of these sequences derive from adult FBFs collected from marine bony fishes (Euteleostei) (Holzer et al., 2008; Nolan and Cribb, 2006a,b). No shark or ray FBF has been sequenced to date, but the holocephalan blood fluke *C. trondheimensis* has been hypothesized as a lineage basal to all other FBFs (Bullard et al., 2008) or as a close relative of "*Sanguinicola* cf. *inermis*" with indeterminate phylogenetic affiliation to the other FBFs (Cribb et al., 2011).

Nolan and Cribb (2006a) used complete ITS2 rDNA to show genetic distance and intraspecific conservation of ITS2 sequences between and within, respectively, select species of *Phthinomita* (*P. littlewoodi*, *P. jonesi*, and *P. hallae*), making the case that ITS2 is a reliable species-level barcode for FBFs. These same authors presented an ITS2 phylogeny for *Phthinomita* spp. and *Ankistromeces* spp. and overlaid the host affiliations for each FBF species. Morphologically similar species of *Phthinomita* infecting siganids were not monophyletic nor was *Phthinomita*. The species of *Ankistromeces* were monophyletic, forming a crown group sister to "*Phthinomita* sp. C" infecting siganids and related to all species of paraphyletic *Phthinomita*. The species of *Ankistromeces* likewise infected not only siganids but also a filefish (Tetraodontiformes: Monacanthidae). Noteworthy is that ITS2 data show that sister species of *Phthinomita* and *Ankistromeces* infected bony fishes that are phylogenetically unrelated but occupy similar niches on the Great Barrier Reef. This perhaps indicates that ecological factors (proximity, abundance, and habitat distribution of invertebrate intermediate hosts) and "host switching" drive FBF speciation in that coral reef system more so than definitive host ancestry. The identities of invertebrate intermediate hosts for species of *Phthinomita* and *Ankistromeces* remain indeterminate. Nolan and Cribb (2006b) used distance analysis (minimum evolution and neighbour joining) of complete ITS2 rDNA sequences to differentiate 4 clades among *Cardicola* and *Braya* on the Great Barrier Reef. These clades agreed with the infected host groups: *Cardicola* spp. infecting rabbitfishes (Siganidae), snappers (Lutjanidae), butterflyfishes (Chaetodontidae), and tunas (Scombridae) and *Braya* spp. infecting parrotfishes (Scaridae) only. Their results demonstrated paraphyly of *Cardicola*, with *Cardicola* spp. infecting tunas comprising a sister clade to that of *Braya* spp. This result was consistent with previous morphology-based studies of *Cardicola* spp. (Bullard, 2010, 2013; Bullard and Overstreet, 2003; Bullard et al., 2012) that indicated a taxonomic revision of the genus was needed concomitant with a reconsideration of *Elaphrobates*, which may not be distinct from *Cardicola*.

ITS2 rDNA sequence data and analysis have yielded inconsistent phylogenetic results. Aiken et al. (2007) used Bayesian inference for complete ITS2 rDNA sequence data for *Pearsonellum corventum* (out-group, parasite of groupers (Serranidae)), *Braya* spp., *C. forsteri*, several innominate species of *Cardicola*, and nine species of *Cardicola* described by Nolan and Cribb (2006b). Their resultant topology, like that of Nolan and Cribb (2006b), revealed that *C. milleri* is sister to all other named non-*C. forsteri* taxa. However, unlike Nolan and Cribb (2006b), Aiken et al. (2007) reported 2 clades comprising (*C. covacinae*(*C. coeptus*(*C. bartolii*, *C. watsonensis*))) and (*C. lafii*, *C. parilus*(*C.* sp. 3(*C. tantabiddii*, *C.* sp. 2))), whereas Nolan and Cribb (2006b) showed (*C. coeptus*(*C. covacinae*(*C. bartolii*, *C. watsonensis*))). Significant also is that the tuna blood flukes *C. forsteri*, "*Cardicola* sp. 3", and "*Cardicola* sp. 4" were monophyletic and sister to the *Braya* spp. in the phylogeny of Nolan and Cribb (2006b), but they were paraphyletic in Aiken et al. (2007), with *Cardicola* spp. that infect mackerels (*Scomberomorus* spp.)

("*Cardicola* sp. 4" and "*Cardicola* sp. 5") sister to tuna blood flukes, snappers (Lutjanidae), and rabbitfishes (Siganidae). Noteworthy here is that, although using the same molecular marker (ITS2) applied to most of the same in-group taxa (except *C. chaetodontis*, not included in Aiken's et al. (2007) analysis), out-group choice seemingly influenced the resultant topology: instead of using *P. corventum* as the out-group, Nolan and Cribb (2006b) used *C. forsteri*. Although results differ, both indicate that *Cardicola* needs revision (op. cit.) and perhaps that distinct FBF genera are justified for snappers (Lutjanidae), with FBFs of spinefoots (Siganidae) harbouring species of a closely related but distinct genus.

Bullard et al. (2008) compared near-complete small-subunit ribosomal DNA sequence data for *C. trondheimensis*, *Acipensericola petersoni*, the putative species of *Sanguinicola* ("*Sanguinicola* cf. *inermis*"), *Aporocotyle spinosicanalis*, *P. acanthus*, and *Neoparacardicola nasonis*. The resultant topology suggested that basal gnathostomes (represented by Chondrichthyes) and basal actinopterygians (represented by Acipenseriformes) harbour lineages of FBFs that are basal to those of higher bony fishes (Euteleostei). This contradicts Smith (1997a), who argued that FBFs lack detectable phylogenetic specificity to particular host lineages. Reports from a growing number of FBF genera indicate some level of phylogenetic host specificity based on both molecular data (Aiken et al., 2007; Holzer et al., 2008; Nolan and Cribb, 2006a,b) and morphology. Simply, closely related blood flukes seemingly infect closely related hosts, for example, blood flukes of lamniform sharks (Orélis-Ribeiro et al., 2013), sturgeon and paddlefish (Acipenseriformes) (Bullard et al., 2008), walking catfishes (Siluriformes: Clariidae) (Truong and Bullard, 2013), drums (Perciformes: Sciaenidae) (Bullard and Overstreet, 2004; Bullard et al., 2012), groupers (Serranidae) (Bullard, 2012; Nolan and Cribb, 2004a; Overstreet and Køie, 1989), and amberjacks (*Seriola* spp.) (Holzer et al., 2008).

Holzer et al. (2008) conducted a phylogenetic analysis of partial 28S rDNA sequences from several putative FBF cercaria (see Brant et al., 2006), some of which were curiously basal to *C. trondheimensis*, and all remaining FBFs analysed. Their analysis showed that the blood flukes (*Paradeontacylix* spp.) of amberjacks (*Seriola* spp.) were monophyletic and formed a clade sister to *C. forsteri* and *Cardicola coeptus*. Their analysis of ITS2 sequence data yielded a phylogeny that left unresolved the relationship between *Psettarium sinensis* (as *Paradeontacylix*), *Braya* spp., and a clade including *Cardicola* spp. and *Paradeontacylix* spp. This phylogeny proved *Cardicola* to be paraphyletic and had a topology for the remaining taxa that was comparable to that of Nolan and Cribb (2006b; see preceding text for discrepancies between Aiken et al. (2007) and these works). Their ITS2 results grouped *Cardicola aurata* as the sister taxon to *Paradeontacylix* (0.54 nodal support).

Cribb et al. (2011) conducted a Bayesian analysis of partial large subunit rDNA sequences across available FBF sequences to establish the taxonomic identity of cercaria isolated from the terebellid polychaete *L. modestus*. Although not the primary purpose of their study, this phylogeny (see Fig. 1 of Cribb et al., 2011) along with the analyses conducted by Bray et al. (2012) (see later text for more details) offers the most phylogenetic breadth previously published for blood flukes, including 52 taxa. That tree is interesting for several reasons. First, Aporocotylidae is paraphyletic, a result that is concordant with the paraphyly of "fishes". Second, *C. trondheimensis* and an assumed species of *Sanguinicola* are in a clade that forms a polytomy with two lineages of blood flukes, including schistosomes and the paraphyletic spirorchiids and marine FBFs. This tree topology potentially foretells the level of taxonomic revision necessary for not only FBFs but also Schistosomatoidea *sensu lato*. Cribb et al. (2011) also detailed records of known and probable FBF cercaria, their hosts, and their morphological features and included a detailed discussion of how much we do not know about the morphology of FBF cercariae. Such discussion echoed that of Brant et al. (2006), who also suggested that diagnostic morphological features of FBF cercaria have been insufficiently explored.

Alama-Bermejo et al. (2011) presented topologies based upon partial 28S rDNA and complete ITS2 rDNA. The former phylogenetic hypothesis comprised three distinct lineages: *P. acanthus*, a clade comprising monophyletic *Cardicola* spp. and *Paradeontacylix* spp., and a clade including *S. meningialis* (an ecologically bizarre FBF that infects the ectomeningeal veins of the brain in common two-banded sea bream *Diplodus vulgaris*), *Psettarium sinensis* (as *Paradeontacylix sinensis*), and *Sasala nolani* (as "Sanguinicolid sp. Moorea-DTJL-2002"). The latter tree showed the two species of *Pearsonellum* as sister to (*S. meningialis*(*Psettarium* sp. KH2007 (*Psettarium* sp. Aburatsubo 3.2 EF544056))).

Ogawa et al. (2011) presented trees based on complete ITS2 rDNA and partial 28S rRNA. The ITS2 analysis (Bayesian) resulted in a topology that showed *Psettarium sinensis* (as *Paradeontacylix sinensis*) and *P. conventum* as two lineages basal to the clade comprising four distinct lineages: *Braya* spp., *Cardicola* spp. ("*Cardicola* 1"), three species of *Cardicola* and monophyletic *Paradeontacylix*, and three species of *Cardicola* that infect tunas

(Scombridae). The latter topology showed a paraphyletic *Cardicola*, a monophyletic *Paradeontacylix*, and *Psettarium sinensis* and *C. orientalis* as distinct lineages.

Bray et al. (2012) conducted a Bayesian analysis assembling three data sets, i.e., 18S, 28S, and concatenated 18S + 28S, to test the phylogenetic position of *Sasala nolani* with other available FBFs sequences. Due to greatest taxonomic coverage, only the 28S tree including 31 taxa was reported, showing a topology concordant with Cribb et al.'s (2011) tree regarding the paraphyly of the family. However, noteworthy is that Bray et al. (2012) reported 2 clades comprising (*N. nasonis*(*P. acanthus*(*S. nolani*, *P. sinensis*(*S. meningialis*)))) and (*Paradeontacylix iberic*us(*P. grandispinus* (*P. balearicus*, *P. kampachi*(*P. godfreyi*(*C. aurata*))))), whereas Cribb et al. (2011) showed (*P. acanthus*(*N. nasonis*(*S. meningialis*(*S. nolani*, *P. sinensis*)))) and (*P. grandispinus*, *P. balearicus*(*P. godfreyi*(*P. kampachi*, *P. ibericus*))), with *C. aurata* in a sister clade, i.e., *Cardicola* clade, grouping with *C. coeptus*, *C. forsteri*, and cercaria from *L. modestus* and *Reterebella aloba*. Such discrepancies may be attributable to the differences in taxon sampling, resulting in differing alignment length and nucleotide substitution models estimated for the data sets. In specific, Cribb et al. (2011) studied an alignment comprising 859 nucleotide bases and TVM + I + G was the model predicted, whereas Bray et al. (2012) considered 693 bases and GTR + I + G was the model chosen.

5. APPROACH TO OUR PHYLOGENETIC ANALYSIS

We analysed GenBank sequence data for the partial D1–D2 domains of 28S and reconstructed a phylogeny that included 88 taxa from Lockyer et al. (2003b), Morgan et al. (2003), Snyder (2004), Brant (2007), Brant and Loker (2009), Hanelt et al. (2009), Tkach et al. (2009), Cribb et al. (2011), and Brant et al. (2012). This phylogeny included new sequences from *Cardicola currani* (KJ572524), *Cardicola palmeri* (KJ572525), and *Elaphrobates euzeti* (KJ572526). Total genomic DNA from newly collected specimens was extracted using DNeasy™ Blood and Tissue kit (QIAGEN) according to the manufacturer's protocol. PCR was carried out using the primer combination U178 + L1642 (Table 1.3) following the method described in Lockyer et al. (2003b). The PCR products were purified using the QIAquick Gel Extraction Kit (QIAGEN) following the manufacturer's protocols. DNA sequencing was performed by GENEWIZ with ABI Prism 3730xl DNA analysers (GENEWIZ, Inc., South Plainfield, NJ) using the same primers as used in the PCR. We thought that the inclusion of additional schistosome and FBF taxa could be helpful because dense taxon sampling can yield more accurate estimates of evolutionary relationships (Heath et al., 2008).

Regarding the out-group, initially, the 28S data set was analysed with an aspidogastrean taxon (*Rugogaster hydrolagi* (AY157176)) and three bivesiculids (*Bivesicula claviformis* (AY222182), *Bivesicula unexpecta* (AY222181), and *Bivesiculoides fusiformis* (AY222183)) along with all available FBF sequences, "Spirorchiidae", and Schistosomatidae plus those of select members of Diplostomidae (*Hysteromorpha triloba* (HM114365) and *Alaria alata* (AF184263)), Strigeidae (*Ichthyocotylurus erraticus* (AY222172)), Clinostomidae (*Clinostomum* spp. (AY222175-6)), and Brachylaimoidea (*Brachylaima virginianum* (DQ060330) and *Brachylaima thompsoni* (AF184262)). In that analysis, the *G. amoena* + *Hapalorhynchus gracilis* clade was the most basal tetrapod blood fluke lineage, i.e., sister to the remaining turtle blood flukes and schistosomes. Previous phylogenetic inferences position that clade sister to a group comprising marine turtle blood flukes plus schistosomes, with blood flukes of freshwater turtles as basal to those lineages, for example, Brant and Loker (2005) and Loker and Brant (2006). Diplostomoidea has previously been considered the sister taxon to all blood flukes (Olson et al., 2003), and, hence, our remaining analyses were performed with Diplostomoidea as the out-group.

Regarding the in-group, previous authors have used unspecified larval clinostomes (AY222175, metacercaria from unspecified site in firetail gudgeon *Hypseleotris galii* (Perciformes: Eleotridae) from coastal streams in Australia, and AY222176, metacercaria from unspecified site in American bullfrog *Rana catesbeiana* from the United States) in blood fluke phylogeny (Bray et al., 2012; Cribb et al., 2011; Olson et al., 2003). The justification for the inclusion of clinostomes in resolving blood fluke phylogeny arises from similarity in sequence data and cercarial morphology. The ssrDNA + lsrDNA phylogeny of Olson et al. (2003) shows a sister group relationship between those unspecified clinostome sequences and that of some FBFs. In addition, and perhaps highly significant, clinostome cercariae have a reportedly uncanny morphological similarity to the blood flukes of fishes and tetrapods (Dönges, 1974; Kirk and Lewis, 1993). In addition, as previously stated earlier in the text, several sequences sourced from unspecified cercarial infections of freshwater gastropods from Eastern Europe, Africa, and Australia similarly have been used (Brant et al., 2006; Olson et al., 2003; discussions in the preceding text). Olson et al. (2003) used sequences derived from cercariae (AY222180) to represent "Sanguinicolidae" that were identified as "*Sanguinicola* cf. *inermis*" because they infected a freshwater gastropod (*Lymnaea stagnalis*) in a water body known to harbour carps (*Cyprinus* spp.) infected by an alleged species of *Sanguinicola* (see Bullard et al., 2008; Kirk, 2012). The original identification of these specimens has drifted, with subsequent authors misrepresenting Olson et al.'s (2003) identification by reporting it as "*Sanguinicola inermis*" or by using AY222180 as a definitive representative of the genus *Sanguinicola* in FBF phylogenetic studies. Nonexistent, however, is published morphological evidence that those specimens were a species of *Sanguinicola* or *S. inermis*, and, to our knowledge, no voucher specimen exists. For this reason, we caution authors in using AY222180 as a definitive representative of *Sanguinicola*. Brant et al. (2006) used phylogenetic inference to identify their cercariae as FBFs (AY585878-81; see discussion earlier in the text) and including AY222180 as "Sanguinicolid sp." Noteworthy here is that, except for the 18S sequence data from adults of *A. petersoni* (see Bullard et al., 2008), nonexistent is sequence data sourced from a definitively identified adult FBF that infects a primary division freshwater fish host. Obviously, molecular phylogenetic inference is a powerful tool in the identification of cercaria; however, it should be understood that we remain unconvinced of the species-, genus-, or family-level identities of these cercariae that have collectively been derived from freshwater gastropods and previously treated as FBFs. This is particularly problematic regarding our understanding of the marine/freshwater origins of blood fluke lineages. Adding to the complexity of the matter, "*Sanguinicola*" traditionally has been used as a repository for any freshwater FBF, and the genus as it is broadly interpreted now likely includes several genera.

To test the effect of including 28S sequence data from species of *Clinostomum* and the freshwater gastropod cercariae of Olson et al. (2003) and Brant et al. (2006), we analysed the data set as +fw gastropod cercariae/−clinostomes (Fig. 1.2A), −fw gastropod cercariae/+clinostomes (Fig. 1.2B), +fw gastropod cercariae/+clinostomes (Fig. 1.2C), and −fw gastropod cercariae/−clinostomes (Fig. 1.2D). Sequences were aligned using MUSCLE version 3.7 (Edgar, 2004) with CLUSTALW sequence weighting and UPGMA clustering for iterations 1 and 2. Resultant alignment was refined by eye using MEGA version 5.2.2 (Tamura et al., 2011) and ends of each fragment were trimmed to match the shortest sequence. Ambiguously aligned positions were identified and removed using a Gblocks server (Castresana, 2000) with all default settings for a less stringent selection. Bayesian inference was performed using MrBayes version 3.2.2 (Huelsenbeck and Ronquist, 2005; Huelsenbeck et al., 2001; Ronquist and Huelsenbeck, 2003) run on CIPRES Portal (Miller et al., 2010). The software jModelTest version 2.1.4 (Darriba et al., 2012; Guindon and Gascuel, 2003) was used to select an appropriate substitution model. The GTR + I + G (proportion of invariable sites = 0.287 and gamma distribution = 1.352) model was inferred as the best estimator by the Akaike information criterion (AIC); therefore, Bayesian analyses used the following parameters: nst = 6, rates = invgamma, ngammacat = 4, and default priors. Analyses were run in duplicate each containing four simultaneous Markov chain Monte Carlo (MCMC) methods (nchains = 4) for 1.0×10^7 generations (ngen = 10,000,000) sampled at intervals of 1000 generations (samplefreq = 1000). Results of the first 3000 trees were discarded as "burn-in" based on the "stationarity" of all parameters sampled by the chains and assessed using Tracer version 1.5 (Rambaut and Drummond, 2009). All retained trees were used to estimate posterior probability of each node. The resulting data matrix for the analysis discussed in the succeeding text (Figs. 1.2A and 1.3) comprised 647 positions per taxon (224 conserved, 423 variable, and 358 parsimony informative), and posterior probability provided strong support to individual nodes.

6. RESULTS FROM OUR PHYLOGENETIC ANALYSIS

The phylogeny based on the inclusion of the freshwater gastropod cercariae and exclusion of clinostomes (Figs. 1.2A and 1.3) forms the foundation for the discussion of the succeeding text. However, we find it noteworthy how the topology changes by including/excluding those sequences (Fig. 1.2B–D). If including the freshwater gastropod cercariae only (Fig. 1.2A), FBFs are monophyletic (all sequences sourced from adult specimens) and the freshwater gastropod cercariae are the sister group to all other blood flukes. If including clinostomes only (Fig. 1.2B), nodal support for a monophyletic Aporocotylidae increases slightly (0.79–0.89) and nodal support for a monophyletic turtle blood flukes plus schistosomes decreases slightly (0.93–0.88). If both the freshwater cercariae and clinostomes are included (Fig. 1.2C), the freshwater gastropod cercariae, marine bony FBFs, and *C. trondheimensis* are recovered as distinct lineages forming a polytomy, with relatively low (0.73) nodal support for the monophyly of clinostomes plus all non-FBFs.

Noteworthy here is that when clinostomes are included, they are the sister group to the blood flukes that infect tetrapods (spirorchiids and schistosomes) and they make more ambiguous the phylogenetic relationship between the freshwater gastropod cercariae and the blood flukes of fishes and tetrapods. On the face of it, some of us (SAB and ROR) initially found it difficult to accept that spirorchiids and schistosomes share a more recent common ancestor with the clinostomes than with the FBFs, i.e., clinostome membership within Schistosomatoidea. Yet, one of us (THC), who is more familiar with clinostomes, finds it not inconceivable that the clinostomid life cycle is an extension of a two-host blood fluke life cycle wherein a vertebrate definitive host consumed the "original" definitive host such that the blood fluke infection remained viable in the predator host. Indeed, it is relatively bizarre that clinostomes infect the oesophagus of their vertebrate hosts, rather than the intestine as most other trematodes do. Moreover, clinostomids, spirorchiids, and schistosomatids have ventral suckers, whereas FBFs lack a ventral sucker. Hence, and taking into account the uncanny morphological similarity of their cercariae (see preceding text), although we accept that adult morphology and definitive host associations make the clinostome + blood fluke association unlikely, our minds are open to novel, future interpretations that may seem strange now.

If freshwater gastropod cercariae and clinostomes both are excluded (Fig. 1.2D), a clear sister group relationship between the monophyletic FBFs and all other blood flukes is recovered with high nodal support for both clades (0.90 and 0.98, respectively). However, we think this latter scenario is oversimplified and dodges one of the most interesting aspects of blood fluke phylogeny: the hitherto unresolved phylogenetic affinities among blood flukes of basal fishes, especially those infecting primary division freshwater fishes, euteleosts, tetrapods, and nontetrapod sarcopterygians. As such, and although the taxonomic identities of Olson et al.'s (2003) and Brant et al.'s (2006) cercariae are indeterminate, including them in the analysis is preferable to us because these cercariae (i) undoubtedly show clear molecular phylogenetic affinities to both turtle blood flukes and FBFs, (ii) were collected from freshwater localities, (iii) infected gastropod species belonging to taxonomic groups known as FBF hosts (although little is known about turtle blood fluke intermediate hosts!), and (iv) may well prove to represent FBFs, possibly species of *Sanguinicola* or novel, closely related genera. We also note the possibility, however distant, that one or several of Brant et al.'s (2006) cercariae from Africa could mature in lungfishes (Sarcopterygii: Dipnoi: Ceratodontiformes: Protopteridae), which would prove particularly exciting since lungfishes are the immediate ancestor to all terrestrial craniates, the sister lineage to Tetrapoda, and presently lack any record of an infection by a blood fluke. Indeed, the localities for these cercarial infections (Brant et al., 2006) comprise rivers harbouring populations of African lungfishes, *Protopterus* spp. (Berra, 2007). Finally, and perhaps not coincidentally, the present parasite phylogeny recovered that places those cercariae as the sister group to the tetrapod blood flukes (Figs. 1.2A and 1.3) mirrors the phylogenetic arrangement of craniates (Fig. 1.2E) regarding Actinopterygii as the sister group to lungfishes and terrestrial craniates (Amemiya et al., 2013; Nelson, 2006), if coelacanths are excluded.

7. SUMMARY OF PHYLOGENETIC STUDY

Considering these aspects, we primarily discuss the tree that includes the freshwater gastropod cercariae along with all other sequences sourced from adult FBFs (Fig. 1.3). Several main observations are worthy of note.

1. With the lowest taxon sampling of all blood flukes in the present analysis, considering only sequences derived from adult specimens, and without sequence data sourced from taxonomically identified adults of any freshwater FBF, the monophyly of Aporocotylidae was not rejected. *C. trondheimensis*, a blood fluke that matures in chimaeras (Chondrichthyes: Holocephali), clearly represents a distantly related lineage that is the sister taxon to the FBFs infecting marine bony fishes (Figs. 1.2A and 1.3). It clusters with FBFs infecting marine bony fishes with a higher posterior probability (0.90) when clinostomes and the freshwater gastropod cercariae were excluded from the analysis. The presence of C-shaped tegumental body spines, unique to species of *Chimaerohemecus*, *Selachohemecus*, and *Hyperandrotrema*, suggests that the blood flukes of chimaeras and sharks are monophyletic. Sequence data from those additional taxa are pending.
2. The blood flukes of bony fishes (Euteleostei) were monophyletic. No 28S sequence data derived from a definitively identified adult FBF infecting a primary division freshwater fish exist to date (see discussion earlier in the text). Hence, by default, all of the FBFs that infect bony fishes and for which 28S sequence data exist are marine species. Nevertheless, the present analysis shows that those taxa are monophyletic. If the clade that includes the unidentified African and Australian cercariae and *Sanguinicola* cf. *inermis* does in fact represent blood flukes that mature in primary division freshwater fishes, then the "monophyly" of the marine

FBFs we show in the present phylogeny would become more meaningful. Indeed, it would also indicate paraphyly of the Aporocotylidae. This gap in sequence data representative of "freshwater FBFs" needs to be closed and should include data from *Sanguinicola*, *Plehniella*, *Nomasanguinicola*, *Acipensericola*, and the other genera of blood flukes that infect primary division freshwater fish lineages. Adult specimens corresponding to those sequences should be lodged in curated helminthological museum. Until then, the discussion of the marine and freshwater monophyly and/or origins of the FBFs is speculative at best. However, insights from comparative morphology have been discussed in detail in a series of systematic works for FBFs that infect primary division freshwater fishes (Bullard et al., 2008; Bullard, 2013; Bullard, in press; ROR and SAB, unpublished observations; Truong and Bullard, 2013). These results together strongly indicate that the blood flukes of primary division freshwater fishes likely share a recent common ancestor and that they exhibit phylogenetic affinities to the turtle blood flukes, suggesting paraphyly of FBFs.

3. Generic interrelationships among the blood flukes of marine bony fishes were consistent with previous phylogenetic studies: *Paradeontacylix* and *Cardicola* were closely related (Holzer et al., 2008), *Cardicola* was not supported as monophyletic and apparently needs taxonomic revision (Bullard, 2013), *Elaphrobates* was not supported as valid (Bullard and Overstreet, 2003; Nolan and Cribb, 2006b), and *Aporocotyle* and *Paradeontacylix* were each monophyletic (see also discussions in the preceding text). No sequence data exist for the type species of *Cardicola* (*Cardicola cardiocolum*), although the present study provides such data for the type and only nominal species of *Elaphrobates*. In fact, few type species have been sequenced, and those that have been sequenced belong to monotypic genera, i.e., *Chimaerohemecus*, *Plethorchis*, *Elaphrobates*, *Skoulekia*, *Sasala*, *Paracardicoloides*, and *Acipensericola* (Fig. 1.1). The only genera that are not monotypic and that have had their type species sequenced comprise *Ankistromeces*, *Braya*, *Pearsonellum*, and *Phthinomita* (Fig. 1.1). This represents another obvious gap in molecular taxonomic data that should be closed in order to test the monophyly of these genera, objectively assess homology, and test the reliability of differential morphological features used in generic diagnoses and species descriptions.
4. Cercariae previously identified as putative FBFs based upon phylogenetic inference (Brant et al., 2006) formed a clade sister to all spirorchiids and schistosomes. However, we remain unconvinced that the cercaria

from *L. stagnalis* (AY222180) is a species of *Sanguinicola*, and the present analysis, although perhaps overly conservative, does not definitively support its membership as a FBF. This sequence remains problematic because its species affiliation was never determined, no voucher material exists, and no sequence data derived from an adult specimen of a species of *Sanguinicola* exist to date (see discussion earlier in the text). If such an affiliation is confirmed, the monophyly of Aporocotylidae would be rejected.

5. All of the known life cycles for freshwater FBFs use gastropods as intermediate hosts (Hoffman et al., 1985; Meade, 1967; Meade and Pratt, 1965; Schell, 1974; Wales, 1958). Similarly, along with the putative cercariae we discussed in the preceding text, all spirorchiids and schistosomes use gastropods as intermediate hosts (Brant et al., 2006). Marine FBFs reportedly use bivalves and polychaetes only, not gastropods (Fraser, 1967; Holliman, 1961; Køie, 1982; Linton, 1915; Martin, 1952; Oglesby, 1961; Wardle, 1979). Considering this ecological similarity, we acknowledge the possibility that the freshwater FBFs likewise have a closer phylogenetic affiliation with spirorchiids and schistosomes than with marine FBFs. We know strikingly little about the life cycles for the majority of named marine and freshwater FBFs (see text earlier) and almost nothing about any marine turtle blood fluke life cycle (see Stacy et al. (2010) for a molecular inference that suggested a limpet (*Fissurella nodosa*) intermediate host). However, the putatively profound dichotomy comprising the freshwater and marine FBFs may reflect this pattern of intermediate host natural history.
6. Regarding the blood flukes that do not infect fishes, i.e., those of tetrapods, our analysis, which used more taxa than any previous analysis, but less sequence data, did not significantly refute the previously published tree topologies for blood flukes, i.e., the monophyly of Schistosomatidae and paraphyly of Spirorchiidae (see Brant and Loker, 2005, 2013; Brant et al., 2006; Lockyer et al., 2003b; Loker and Brant, 2006; Snyder, 2004; Snyder and Loker, 2000).
7. Regarding schistosome interrelationships, the resulting tree generally matched those recovered by previous workers (Brant and Loker, 2005; Brant et al., 2006; Lockyer et al., 2003b; Loker and Brant, 2006; Snyder, 2004; Snyder and Loker, 2000): (i) high support for clades AO (*Austrobilharzia*, *Ornithobilharzia*), BSO (*Bivitellobilharzia*, *Schistosoma*, *Orientobilharzia*), SH (*Schistosomatium*, *Heterobilharzia*), and BTGB (*Bilharziella*, *Trichobilharzia*, *Dendritobilharzia*, *Gigantobilharzia*);

(ii) clade AO basal; (iii) clade SH sister to *Dendritobilharzia* + *Bilharziella* + *Gigantobilharzia* + *Trichobilharzia* + *Allobilharzia* + *Anserobilharzia*; (iv) *Bivitellobilharzia* sister to clade SO; and (v) Asian schistosomes basal (Brant and Loker, 2005, 2009, 2013; Brant et al., 2006; Lawton et al., 2011; Lockyer et al., 2003b; Loker and Brant, 2006; Snyder, 2004; Snyder and Loker, 2000; Wang et al., 2009; Webster and Littlewood, 2012; Webster et al., 2006).

8. Although not the focus of the work presented here, we note that some schistosome interrelationships were unlike those previously recovered, perhaps because of the proportionally smaller amount of molecular sequence data that were included in our analysis: (i) *Dendritobilharzia* basal to (*Gigantobilharzia* (*Bilharziella* + *Trichobilharzia*)) and (ii) *Macrobilharzia*, *Bivitellobilharzia*, SO, and SH + BTGB unresolved. Moreover, the present study indicated that mammal schistosomes are monophyletic, indicating three independent colonizations of birds. Noteworthy also is that *G. amoena* was sister to the freshwater turtle blood flukes, apparently not wholly supporting the notion that schistosomes colonized modern archosaurs (birds) from ancestral archosaurs (crocodilians), i.e., *G. amoena*, although reported dioecious, is not a member of Schistosomatidae (see Brant and Loker, 2005; Brant et al., 2013; and Loker and Brant, 2006).

8. FUTURE DIRECTIONS

The Digenea is perhaps the largest group of endoparasitic metazoan parasites and includes 150 families with 2700 nominal genera, >18,000 nominal species, and conservatively ∼40,000 extant species (Cribb et al., 2001). The ubiquity of FBFs reveals them as likely significant actors in freshwater, marine, and estuarine ecosystems (Cribb et al., 2001). However, limited taxon sampling in FBF phylogenetic studies has hampered understanding of coevolution with host taxa and placement of "Aporocotylidae" within Schistosomatoidea. We think that an understanding of the evolutionary origins of flatworm parasitism in the blood of craniates will be advanced significantly by continued studies on the morphology and molecular biology of blood flukes that infect basal fishes, for example, Chondrichthyes, Acipenseriformes, Elopiformes, and Siluriformes. As with the sarcopterygian fish lineages, coelacanths and lungfishes, we also emphasize the need to examine hagfishes (Myxiniformes) and lampreys (Petromyzontiformes) for the presence of blood fluke infections. To our knowledge, no such examinations have been conducted, but finding and describing blood fluke specimens from these fascinating and phylogenetically unique craniates would be exciting and impactful to our understanding of the evolution of blood flukes as a whole. Hagfishes and lampreys may well harbour novel genera and/or families of blood flukes. Such increased taxon sampling for morphological features and molecular sequence data and the development and application of novel molecular markers should help resolve the interrelationships of the FBFs, analogous to the resolution of the paraphyletic "Spirorchiidae" that included 18S rDNA and 28S rDNA sequences from freshwater and marine representatives of eight spirorchiid genera (Snyder, 2004). Moreover, a comprehensive molecular phylogeny of blood flukes underpins testing hypotheses about the origin of diseases caused by blood flukes in craniates and trematode dioecity (Platt and Brooks, 1997), a condition that is unusual among flatworms and may have independently evolved several times in the Digenea.

A substantial quantity of genomic information, which exists for schistosomes, is lacking and wholly unexplored in nonhuman blood flukes, especially among FBFs. Massive parallel sequencing platforms, i.e., next- or second-generation sequencing, are opening new avenues to life sciences research by high-throughput technology that can inexpensively and within days produce thousands of megabases of nucleotide information. The commercially available platforms are distinguished by a combination of specific protocols that can be arranged as template preparation (clonally amplified or single-molecule templates), sequencing and imaging, and data analysis (Metzker, 2010). Such technology recently has improved the draft genome of *Schistosoma mansoni*, which is not surprising given its medical importance, and hastened completion of its genome, i.e., the first among parasitic flatworms (Berriman et al., 2009; Protasio et al., 2012; Tsai et al., 2013). The study of the evolution of complex body plans, parasite–host relationships, parasite sensory systems, disease pathogenesis, and new drugs and vaccine sites has been explored by the use of molecular tools applied to genome studies in *S. haematobium*, *S. mansoni*, and *S. japonicum* (see Berriman et al., 2009; The *Schistosoma japonicum* Genome Sequencing and Functional Analysis Consortium, 2009; Young et al., 2012). Analogous questions could be applied similarly to FBF studies. Second-generation sequencing technologies have provided high-resolution maps of temporal changes in gene expression among cercaria, schistosomula, and adult stages of *Schistosoma* spp. (Protasio et al., 2012). Similarly, Collins et al. (2013) recently used genomic resources and RNA-seq-based gene expression profiling to identify the gene responsible for the maintenance of neoblast-like cells. These data are directly relevant to a molecular understanding of the way in which these human pathogens infect the host, and they are relevant because blocking infection means preventing disease. These processes likely originated in, and likely remain to be elucidated in, FBF–fish relationships, underscoring the need for genome studies comprising FBFs that infect the major nontetrapod craniate lineages.

ACKNOWLEDGEMENTS

We thank Matthew R. Womble (SAB's laboratory) for assisting in the collection of FBFs from the Gulf of Mexico and Johanna T. Cannon, Dr. Kevin M. Kocot, and Dr. Pamela M. Brannock (all KMH's laboratory), and Dr. Zhen Tao (CRA's laboratory) for technical assistance to R. O. R. This is a contribution of the Southeastern Cooperative Fish Parasite and Disease Project (School of Fisheries, Aquaculture, and Aquatic Sciences, College of Agriculture, Auburn University) and was supported in part by the National Science Foundation's Division of Environmental Biology with funds from NSF-DEB Grant numbers 1112729, 1051106, and 1048523 to S. A. B.

REFERENCES

Aiken, H.M., Bott, N.J., Mladineo, I., Montero, F.E., Nowak, B.F., Hayward, C.J., 2007. Molecular evidence for cosmopolitan distribution of platyhelminth parasites of tunas (*Thunnus* spp.). Fish Fish. 8, 167–180.

Alama-Bermejo, G., Montero, F.E., Raga, J.A., Holzer, A.S., 2011. *Skoulekia meningialis* n. gen., n. sp. (Digenea: Aporocotylidae Odhner, 1912) a parasite surrounding the brain of the Mediterranean common two-banded seabream *Diplodus vulgaris* (Geoffroy Saint-Hilaire, 1817) (Teleostei: Sparidae): description, molecular phylogeny, habitat and pathology. Parasitol. Int. 60, 34–44.

Aldhoun, J.A., Littlewood, D.T.J., 2012. *Orientobilharzia* Dutt & Srivastava, 1955 (Trematoda: Schistosomatidae), a junior synonym of *Schistosoma* Weinland, 1858. Syst. Parasitol. 82, 81–88.

Amemiya, C.T., Alföldi, J., Lee, A.P., Fan, S., Philippe, H., MacCallum, I., Braasch, I., Manousaki, T., Schneider, I., Rohner, N., Organ, C., Chalopin, D., Smith, J.J., Robinson, M., Dorrington, R.A., Gerdol, M., Aken, B., Biscotti, M.A., Barucca, M., Baurain, D., Berlin, A.M., Blatch, G.L., Buonocore, F., Burmester, T., Campbell, M.S., Canapa, A., Cannon, J.P., Christoffels, A., De Moro, G., Edkins, A.L., Fan, L., Fausto, A.M., Feiner, N., Forconi, M., Gamieldien, J., Gnerre, S., Gnirke, A., Goldstone, J.V., Haerty, W., Hahn, M.E., Hesse, U., Hoffmann, S., Johnson, J., Karchner, S.I., Kuraku, S., Lara, M., Levin, J.Z., Litman, G.W., Mauceli, E., Miyake, T., Mueller, M.G., Nelson, D.R., Nitsche, A., Olmo, E., Ota, T., Pallavicini, A., Panji, S., Picone, B., Ponting, C.P., Prohaska, S.J., Przybylski, D., Saha, N.R., Ravi, V., Ribeiro, F.J., Sauka-Spengler, T., Scapigliati, G., Searle, S.M. J., Sharpe, T., Simakov, O., Stadler, P.F., Stegeman, J.J., Sumiyama, K., Tabbaa, D., Tafer, H., Turner-Maier, J., van Heusden, P., White, S., Williams, L., Yandell, M., Brinkmann, H., Volff, J.-N., Tabin, C.J., Shubin, N., Schartl, M., Jaffe, D.B., Postlethwait, J.H., Venkatesh, B., Di Palma, F., Lander, E.S., Meyer, A., Lindblad-Toh, K., 2013. The African coelacanth genome provides insights into tetrapod evolution. Nature 496, 311–316.

Anderson, G.R., Barker, S.C., 1998. Inference of phylogeny and taxonomy within the Didymozoidae (Digenea) from the second internal transcribed spacer (ITS2) of ribosomal DNA. Syst. Parasitol. 41, 87–94.

Attwood, S.W., Panasoponkul, C., Upatham, E.S., Meng, X.H., Southgate, V.R., 2002. *Schistosoma ovuncatum* n. sp. (Digenea: Schistosomatidae) from northwest Thailand and the historical biogeography of Southeast Asian Schistosoma Weinland, 1858. Syst. Parasitol. 51, 1–19.

Barker, S.C., Blair, D., 1996. Molecular phylogeny of *Schistosoma* species supports traditional groupings within the genus. J. Parasitol. 82, 292–298.

Basch, P.F., 1991. Schistosomes: Development, Reproduction and Host Relations, first ed. Oxford University Press, Oxford, UK.

Bazikalova, A., 1932. Beiträge zur Parasitologie der Murman'schen Fische (in Russian). In: Mittelman, S.Y. (Ed.), Sbornik Nauchno-Promyslovikh Rabot na Murman. Narkomsnab SSR Tsentral'nya Institut Rybnogo Khozyaistva, Moskva, Leningrad, pp. 136–153.

Berra, T.M., 2007. Freshwater Fish Distribution. University of Chicago Press, Chicago, USA, 606 pp.

Berriman, M., Haas, B.J., LoVerde, P.T., Wilson, R.A., Dillon, G.P., Cerqueira, G.C., Mashiyama, S.T., Al-Lazikani, B., Andrade, L.F., Ashton, P.D., Aslett, M.A., Bartholomeu, D.C., Blandin, G., Caffrey, C.R., Coghlan, A., Coulson, R., Day, T.A., Delcher, A., Demarco, R., Djikeng, A., Eyre, T., Gamble, J.A., Ghedin, E., Gu, Y., Hertz-Fowler, C., Hirai, H., Hirai, Y., Houston, R., Ivens, A., Johnston, D.A., Lacerda, D., Macedo, C.D., Mcveigh, P., Ning, Z., Oliveira, G., Overington, J.P., Parkhill, J., Pertea, M., Pierce, R.J., Protasio, A.V., Quail, M.A., Rajandream, M.-A., Rogers, J., Sajid, M., Salzberg, S.L., Stanke, M., Tivey, A.R., White, O., Williams, D.L., Wortman, J., Wu, W., Zamanian, M., Zerlotini, A., Frase-Liggett, C.M., Barrell, B.G., El-Sayed, N.M., 2009. The genome of the blood fluke *Schistosoma mansoni*. Nature 460, 352–358.

Bowles, J., Hope, M., Tiu, W.U., Liu, X., Mcmanus, D.P., 1993. Nuclear and mitochondrial genetic markers highly conserved between Chinese and Philippine Schistosoma japonicum. Acta Trop. 55, 217–229.

Brant, S.V., 2007. The occurrence of the avian schistosome *Allobilharzia visceralis* Kolářová, Rudolfová, Hampl et Skirnisson, 2006 (Schistosomatidae) in the tundra swan, *Cygnus columbianus* (Anatidae), from North America. Folia Parasitol. 54, 99–104.

Brant, S.V., Loker, E.S., 2005. Can specialized pathogens colonize distantly related hosts? Schistosome evolution as a case study. PLoS Pathog. 1, 0167–0169.

Brant, S.V., Loker, E.S., 2009. Systematics of the avian schistosome genus *Trichobilharzia* (Trematoda: Schistosomatidae) in North America. J. Parasitol. 95, 941–963.

Brant, S.V., Loker, E.S., 2013. Discovery-based studies of schistosome diversity stimulate new hypotheses about parasite biology. Trends Parasitol. 29, 449–459.

Brant, S.V., Morgan, J.A., Mkoji, G.M., Snyder, S.D., Rajapakse, R.P., Loker, E.S., 2006. An approach to revealing blood fluke life cycles, taxonomy, and diversity: provision of key reference data including DNA sequence from single life cycle stages. J. Parasitol. 92, 77–88.

Brant, S.V., Pomajbikova, K., Modry, D., Petrzelkova, K., Loker, E.S., 2012. Molecular phylogeny of the elephant schistosome, *Bivitellobilharzia loxodontae* (Trematoda: Schistosomatidae) from the Central African Republic. J. Helminthol. 87, 102–107.

Brant, S.V., Jouet, D., Ferte, H., Loker, E.S., 2013. *Anserobilharzia* gen. n. (Digenea, Schistosomatidae) and redescription of *A. brantae* (Farr & Blankemeyer, 1956) comb. n. (syn. *Trichobilharzia brantae*), a parasite of geese (Anseriformes). Zootaxa 3670, 193–206.

Bray, R.A., Cribb, T.H., Littlewood, D.T.J., 2012. *Sasala nolani* gen. n., sp. n. (Digenea: Aporocotylidae) from the body-cavity of the guineafowl puffer fish *Arothron meleagris* (Lacepède) (Tetraodontiformes: Tetraodontidae) from off Moorea, French Polynesia. Zootaxa 3334, 29–41.

Bullard, S.A., 2010. A new species of *Cardicola* Short, 1953 (Digenea: Aporocotylidae) from the heart and branchial vessels of two surfperches (Embiotocidae) in the Eastern Pacific Ocean off California. J. Parasitol. 96, 382–388.

Bullard, S.A., 2012. *Pearsonellum lemusi* n. sp. (Digenea: Aporocotylidae) from blood vascular system of gag grouper, *Mycteroperca microlepis*, (Perciformes: Serranidae) off Alabama, with an emendation of *Pearsonellum* Overstreet and Koie, 1989. J. Parasitol. 98, 323–327.

Bullard, S.A., 2013. *Cardicola langeli* n. sp. (Digenea: Aporocotylidae) from heart of sheepshead, *Archosargus probatocephalus*, (Actinopterygii: Sparidae) in the Gulf of Mexico, with an updated list of hosts, infection sites and localities for *Cardicola* spp. Folia Parasitol. 60, 17–27.

Bullard, S.A., in press. Blood flukes (Digenea: Aporocotylidae) of Elopomorpha: emended diagnosis of *Paracardicoloides*, supplemental observations of *Paracardicoloides yamagutii*, and a new genus and species from ladyfish, *Elops saurus* in the northern Gulf of Mexico.

Bullard, S.A., Jensen, K., 2008. Blood flukes (Digenea: Aporocotylidae) of stingrays (Myliobatiformes: Dasyatidae): *Orchispirium heterovitellatum* from *Himantura imbricata* in the Bay of Bengal and a new genus and species from *Dasyatis sabina* in the Northern Gulf of Mexico. J. Parasitol. 94, 1,311–1,321.

Bullard, S.A., Overstreet, R.M., 2002. Potential pathological effects of blood flukes (Digenea: Sanguinicolidae) on pen-reared marine fishes. Proc. Gulf and Caribb. Fish. Inst. 53, 10–25.

Bullard, S.A., Overstreet, R.M., 2003. *Elaphrobates euzeti* gen. and sp. n. (Digenea: Sanguinicolidae) from snappers (Lutjanidae) in the Gulf of Mexico. In: Combes, C., Jourdane, J. (Eds.), Taxonomie, écologie et évolution des métazoaires parasites. Taxonomy, Ecology and Evolution of Metazoan Parasites, Tome 1. PUP, Perpignan, pp. 97–113.

Bullard, S.A., Overstreet, R.M., 2004. Two new species of *Cardicola* (Digenea: Sanguinicolidae) in drums (Sciaenidae) from Mississippi and Louisiana. J. Parasitol. 90, 128–136.

Bullard, S.A., Overstreet, R.M., 2008. Digeneans as enemies of fishes. In: Eiras, J., Segner, H., Wahil, T., Kapoor, B.G. (Eds.), Fish Diseases. Science Publishers, US, pp. 817–976.

Bullard, S.A., Overstreet, R.M., Carlson, J.K., 2006. *Selachohemecus benzi* n. sp. (Digenea: Sanguinicolidae) from the blacktip shark *Carcharhinus limbatus* in the Northern Gulf of Mexico. Syst. Parasitol. 63, 143–154.

Bullard, S.A., Snyder, S.D., Jensen, K., Overstreet, R.M., 2008. New genus and species of Aporocotylidae (Digenea) from a basal actinopterygian, the American paddlefish, *Polyodon spathula* (Acipenseriformes: Polyodontidae) from the Mississippi Delta. J. Parasitol. 94, 487–495.

Bullard, S.A., Jensen, K., Overstreet, R.M., 2009. Historical account of the two family-group names in use for the single accepted family comprising the "fish blood flukes" Acta Parasitol. 54, 78–84.

Bullard, S.A., Olivares-Fuster, O., Benz, G.W., Arias, C.R., 2011. Molecules infer origins of ectoparasite infracommunities on tunas. Parasitol. Int. 60, 447–451.

Bullard, S.A., Baker, T., de Buron, I., 2012. New species of *Cardicola* Short, 1953 (Digenea: Aporocotylidae) from heart of Atlantic croaker, *Micropogonias undulatus*, (Sciaenidae) of the South Atlantic Bight. J. Parasitol. 98, 328–332.

Castresana, J., 2000. Selection of conserved blocks from multiple alignments for their use in phylogenetic analysis. Mol. Biol. Evol. 17, 540–552.

Chen, C.M., Wen, J.X., Hong, Q., Wang, J.L., Wang, Y.Y., 2008. Primary study on molecular phylogeny of *Paradeontacylix sinensis* Liu (in Chinese). J. Fujian Normal Univ. 24, 71–75.

Collins III, J., Wang, B., Lambrus, B., Tharp, M., Iyer, H., Newmark, P., 2013. Adult somatic stem cells in the human parasite *Schistosoma mansoni*. Nature 494, 476–479.

Cribb, T.H., Anderson, G.R., Adlard, R.D., Bray, R.A., 1998. A DNA-based demonstration of a three-host life-cycle for the Bivesiculidae (Platyhelminthes: Digenea). Int. J. Parasitol. 28, 1791–1795.

Cribb, T.H., Bray, R.A., 2011. Trematode families and genera: have we found them all? Trends Parasitol. 27, 149–154.

Cribb, T.H., Bray, R.A., Littlewood, D.T.J., Pichelin, S., Herniou, E.A., 2001. The Digenea. In: Littlewood, D.T.J., Bray, R.A. (Eds.), Interrelationships of the Platyhelminthes. Taylor and Francis, London, UK, pp. 168–185.

Cribb, T.H., Adlard, R.D., Hayward, C.J., Bott, N.J., Ellis, D., Evans, D., Nowak, B.F., 2011. The life cycle of *Cardicola forsteri* (Trematoda: Aporocotylidae), a pathogen of ranched southern bluefin tuna, *Thunnus maccoyii*. Int. J. Parasitol. 41, 861–870.

Darriba, D., Taboada, G.L., Doallo, R., Posada, D., 2012. jModelTest 2: more models, new heuristics and parallel computing. Nat. Methods 9, 772.

Dönges, J., 1974. The life cycle of *Euclinostomum heterostomum* (Rudolphi, 1809) (Trematoda: Clinostomatidae). Int. J. Parasitol. 4, 79–90.

Edgar, R.C., 2004. MUSCLE: multiple sequence alignment with high accuracy and high throughput. Nucleic Acids Res. 32, 1792–1797.

Fraser, T.H., 1967. Contributions to the biology of *Tagelus divisus* (Tellinacea: Pelecypoda) in Biscayne Bay, Florida. Bull. Mar. Sci. 17, 111–132.

Goto, S., Ozaki, Y., 1929. Brief notes on new trematodes II. Jpn. J. Zool. 2, 369–381.

Guindon, S., Gascuel, O., 2003. A simple, fast and accurate method to estimate large phylogenies by maximum-likelihood. Syst. Biol. 52, 696–704.

Hanelt, B., Brant, S.V., Steinauer, M.L., Maina, G.M., Kinuthia, J.M., Agola, L.E., Mwangi, I.N., Mungai, B.N., Mutuku, M.W., Mkoji, J.M., Loker, E.S., 2009. *Schistosoma kisumuensis* n. sp. (Digenea: Schistosomatidae) from murid rodents in the Lake Victoria Basin, Kenya and its phylogenetic position within the *S. haematobium* species group. Parasitology 136, 987–1001.

Hardy-Smith, P., Ellis, D., Humphrey, J., Evans, M., Evans, D., Rough, K., Valdenegro, V., Nowak, B., 2012. *In vitro* and *in vivo* efficacy of anthelmintic compounds against blood fluke (*Cardicola forsteri*). Aquaculture 334–337, 39–44.

Hayward, C.J., Ellis, D., Foote, D., Wilkinson, R.J., Crosbie, P.B.B., Bott, N.J., Nowah, B.F., 2010. Concurrent epizootic hyperinfections of sea lice (predominantly *Caligus chiastos*) and blood flukes (*Cardicola foresteri*) in ranched Southern Bluefin tuna. Vet. Parasitol. 173, 107–115.

Heath, T.A., Hedtke, S.M., Hillis, D.M., 2008. Taxon sampling and the accuracy of phylogenetic analyses. J. Syst. Evol. 46, 239–257.

Hernández-Orts, J.S., Alama-Bermejo, G., Carrillo, J.M., García, N.A., Crespo, E.A., Raga, J.A., Montero, F.E., 2012. *Aporocotyle mariachristinae* n. sp., and *A. ymakara* Villalba & Fernández, 1986 (Digenea: Aporocotylidae) of the pink cusk-eel, *Genypterus blacodes* (Ophidiiformes: Ophidiidae) from Patagonia, Argentina. Parasite 19, 319–330.

Hoffman, G.L., Freid, B., Harvey, J.E., 1985. *Sanguinicola fontinalis* sp. nov. (Digenea: Sanguinicolidae): a blood parasite of brook trout, *Salvelinus fontinalis* (Mitchill), and longnose dace, *Rhinichthys cataractae* (Valenciennes). J. Fish Dis. 8, 529–538.

Holliman, R.B., 1961. Larval trematodes from the Apalachee Bay area, Florida, with a checklist of known marine cercariae arranged in a key to their superfamilies. Tulane Stud. Zool. 9, 1–74.

Holzer, A.S., Montero, F.E., Repullés, A., Nolan, M.J., Sitja-Bobadilla, A., Alvarez-Pellitero, P., Zarza, C., Raga, J.A., 2008. *Cardicola aurata* sp. n. (Digenea: Sanguinicolidae) from Mediterranean *Sparus aurata* L. (Teleostei: Sparidae) and its unexpected phylogenetic relationship with *Paradeontacylix* McIntosh, 1934. Parasitol. Int. 57, 472–482.

Horák, P., Kolářová, L., Dvorák, J., 1998. *Trichobilharzia regenti* n. sp. (Schistosomatidae, Bilharziellinae), a new nasal schistosome from Europe. Parasite 5, 349–357.

Huelsenbeck, J.P., Ronquist, R., 2005. Bayesian analysis of molecular evolution using MrBayes. In: Nielsen, R. (Ed.), Statistical Methods in Molecular Evolution. Springer-Verlag, New York, USA, pp. 183–232.

Huelsenbeck, J.P., Ronquist, F., Nielsen, R., Bollback, J.P., 2001. Bayesian inference of phylogeny and its impact on evolutionary biology. Science 294, 2310–2314.

Ishimaru, K., Mine, R., Shirakashi, S., Kaneko, E., Kazushige, K., Okada, T., Sawada, Y., Ogawa, K., 2013. Praziquantel treatment against *Cardicola* blood flukes: determination of the minimal effective does and pharmacokinetics in juvenile Pacific bluefin tuna. Aquaculture 2013, 402–403.

Khalil, L.F., 2002. Family Schistosomatidae Stiles & Hassall, 1898. In: Gibson, D.I., Jones, A., Bray, R.A. (Eds.), Keys to the Trematoda. CABI Publishing, Wallingford, UK, pp. 419–432.

Kirchhoff, N.T., Leef, M.J., Valdenegro, V., Hayward, C.J., Nowak, B.F., 2012. Correlation of humoral immune response in Southern bluefin tuna, *T. maccoyii*, with infection stage of the blood fluke, *Cardicola forsteri*. PLoS One 7, e45742.

Kirk, R.S., 2012. 16. *Sanguinicola inermis* and related species. In: Woo, P.T.K., Buchmann, K. (Eds.), Fish Parasites: Pathobiology and Protection. CABI, London, UK, pp. 270–281.

Kirk, R.S., Lewis, J.W., 1993. The life-cycle and morphology of *Sanguinicola inermis* Plehn, 1905 (Digenea: Sanguinicolidae). Syst. Parasitol. 25, 125–133.

Køie, M., 1982. The redia, cercaria and early stages of *Aporocotyle simplex* Odhner, 1900 (Sanguinicolidae)—a digenetic trematode which has a polychaete annelid as the only intermediate host. Ophelia 21, 115–145.

Køie, M., Petersen, M.E., 1988. A new annelid intermediate host (*Lanassa nordenskioeldi* Malmgren, 1866) (Polychaeta: Terebellidae) for *Aporocotyle* sp. and a new final host family (Pisces: Bothidae) for *Aporocotyle simplex* Odhner, 1900 (Digenea: Sanguinicolidae). J. Parasitol. 74, 499–502.

Kolářová, L., Rudolfová, J., Hampl, V., Skirnisson, K., 2006. *Allobilharzia visceralis* gen. nov. sp. nov. (Schistosomatidae—Trematoda) from *Cygnus cygnus* (L.) (Anatidae). Parasitol. Int. 55, 179–186.

Kolářová, L., Skirnisson, K., Ferte, H., Jouet, D., 2013. *Trichobilharzia mergi* sp. nov. (Trematoda: Digenea: Schistosomatidae), a visceral schistosome of *Mergus serrator* (L.) (Aves: Anatidae). Parasitol. Int. 62, 300–308.

Lawton, S.P., Hirai, H., Ironside, J.E., Johnston, D.A., Rollinson, D., 2011. Genomes and geography: genomic insights into the evolution and phylogeography of the genus *Schistosoma*. Parasit. Vectors 4, 131.

Lin, D., Hanson, L.A., Pote, L.M., 1999. Small subunit ribosomal RNA sequence of Henneguya exilis (Class Myxosporea) identifies the actinosporean stage from an oligochaete host. J. Euk. Microbiol. 46, 66–68.

Linton, E., 1915. Sporocysts in an annelid. Biol. Bull. 28, 115–118.

Littlewood, D.T.J., 1994. Molecular phylogenetics of cupped oysters based on partial 28S rRNA gene sequences. Mol. Phylogenet. Evol. 3, 221–229.

Littlewood, D.T.J., Johnston, D.A., 1995. Molecular phylogenetics of the four Schistosoma species groups determined with partial 285 ribosomal RNA gene sequences. Parasitology 111, 167–175.

Littlewood, D.T.J., Olson, P.D., 2001. Small subunit rDNA and the phylum Platyhelminthes: signal, noise, conflict and compromise. In: Littlewood, D.T.J.,

Bray, R.A. (Eds.), Interrelationships of the Platyhelminthes. Taylor & Francis, London, UK, pp. 262–278.

Littlewood, D.T.J., Rohde, K., Clough, K.A., 1997. Parasite speciation within or between host species? Phylogenetic evidence from site-specific polystome monogeneans. Int. J. Parasitol. 27, 1289–1297.

Littlewood, D.T.J., Rohde, K., Clough, K.A., 1999. The interrelationships of all major groups of Platyhelminthes: phylogenetic evidence from morphology and molecules. Biol. J. Linn. Soc. 66, 75–114.

Littlewood, D.T.J., Curini-Galletti, M., Herniou, E.A., 2000. The interrelationships of Proseriata (Platyhelminthes: Seriata) tested with molecules and morphology. Mol. Phylogenet. Evol. 16, 449–466.

Lockyer, A.E., Olson, P.D., Littlewood, D.T.J., 2003a. Utility of complete large and small subunit rRNA genes in resolving the phylogeny of the Neodermata: implications and a review of the cercomer theory. Biol. J. Linn. Soc. 78, 155–171.

Lockyer, A.E., Olson, P.D., Ostergaard, P., Rollinson, D., Johnston, D.A., Attwood, S.W., Southgate, V.R., Horak, P., Snyder, S.D., Le, T.H., Agatsuma, T., Mcmanus, D.P., Carmichael, A.C., Naem, S., Littlewood, D.T.J., 2003b. The phylogeny of the Schistosomatidae based on three genes with emphasis on the interrelationships of *Schistosoma* Weinland, 1858. Parasitology 126, 203–224.

Loker, E.S., Brant, S.V., 2006. Diversification, dioecy and dimorphism in schistosomes. Trends Parasitol. 22, 521–528.

Madhavi, R., Rao, H.K., 1970. *Orchispirium heterovitellatum* gen. et sp. nov. (Trematoda: Sanguinicolidae) from the ray fish, *Dasyatis imbricatus* Day, from Bay of Bengal. J. Parasitol. 56, 41–43.

Manter, H.W., 1940. Digenetic trematodes of fishes from the Galapagos Islands and the neighboring Pacific. Allan Hancock Pac. Exp. 2, 329–497.

Martin, W.E., 1952. Another annelid first intermediate host of a digenetic trematode. J. Parasitol. 38, 1–4.

Martin, W.E., 1960. Hawaiian helminths. IV. *Paracardicola hawaiensis* n. gen., n. sp. (Trematoda: Sanguinicolidae) from the balloon fish, *Tetraodon hispidus* L. J. Parasitol. 46, 648–650.

Meade, T.G., 1967. Life history studies on *Cardicola klamathensis* (Wales, 1958) Meade and Pratt, 1965 (Trematoda: Sanguinicolidae). Proc. Helminthol. Soc. Wash. 34, 210–212.

Meade, T.G., Pratt, I., 1965. Description and life history of *Cardicola alseae* sp. n. (Trematoda: Sanguinicolidae). J. Parasitol. 51, 575–578.

Medlin, L., Elwood, H.J., Stickel, S., Sogin, M.L., 1988. The characterization of enzymatically amplified eukaryotic 16S-like rRNA-coding regions. Gene 71, 491–499.

Metzker, M.L., 2010. Sequencing technologies—the next generation. Nat. Rev. Genet. 11, 31–46.

Miller, M.A., Pfeiffer, W., Schwartz, T., 2010. Creating the CIPRES Science Gateway for inference of large phylogenetic trees. In: Proceedings of Gateway Computing Environments Workshop (GCE), 14 November 2010, New Orleans, LA, pp. 1–8.

Morgan, J.A.T., Dejong, R.J., Kazibwe, F., Mkoji, G.M., Loker, E.S., 2003. A newly identified lineage of *Schistosoma*. Int. J. Parasitol. 33, 977–985.

Müller, V., Kimmig, P., 1994. *Trichobilharzia franki* n. sp.—the cause of swimmer's dermatitis in Southwest German dredges lakes. Appl. Parasitol. 35, 12–31.

Nelson, J.S., 2006. Fishes of the World, fourth ed. John Wiley and Sons, Inc., New York, NY, USA

Nolan, M.J., Cribb, T.H., 2004a. The life cycle of *Paracardicoloides yamagutii* Martin, 1974 (Digenea: Sanguinicolidae). Folia Parasitol. 51, 320–326.

Nolan, M.J., Cribb, T.H., 2004b. Two new blood flukes (Digenea: Sanguinicolidae) from Epinephelinae (Perciformes: Serranidae) of the Pacific Ocean. Parasitol. Int. 53, 327–335.

Nolan, M.J., Cribb, T.H., 2004c. Ankistromeces mariae n. g., n. sp. (Digenea: Sanguinciolidae) from *Meuschenia freycineti* (Monacanthidae) off Tasmania. Syst. Parasitol. 57, 151–157.

Nolan, M.J., Cribb, T.H., 2005. The use and implications of ribosomal DNA sequencing for the discrimination of digenean species. Adv. Parasitol. 60, 101–163.

Nolan, M.J., Cribb, T.H., 2006a. An exceptionally rich complex of Sanguinicolidae von Graff, 1907 (Platyhelminthes: Trematoda) from Siganidae, Labridae and Mullidae (Teleostei: Perciformes) from the Indo-west Pacific Region. Zootaxa 1218, 1–80.

Nolan, M.J., Cribb, T.H., 2006b. *Cardicola* Short, 1953 and *Braya* n. gen. (Digenea: Sanguinicolidae) from five families of tropical Indo-Pacific fishes. Zootaxa 1265, 1–80.

Norte Do Santos, C.C., Leef, M.J., Jones, J.B., Bott, N.J., Giblot-Ducray, D., Nowak, B., 2012. Distribution of *Cardicola forsteri* eggs in the gills of Southern bluefin tuna (*Thunnus maccoyii*) (Castelnau, 1872). Aquaculture 344–349, 54–57.

Ogawa, K., Nagano, T., Akai, N., Sugita, A., Hall, K.A., 2007. Blood fluke infection of cultured tiger Puffer *Takifugu rubripes* imported from China to Japan. Fish Pathol. 42, 91–99.

Ogawa, K., Tanaka, S., Sugihara, Y., Takami, I., 2010. A new blood fluke of the genus *Cardicola* (Trematoda: Sanguinicolidae) from Pacific bluefin tuna *Thunnus orientalis* (Temminck & Schlegel, 1844) cultured in Japan. Parasitol. Int. 59, 44–48.

Ogawa, K., Ishimaru, K., Shirakashi, S., Takami, I., Grabner, D., 2011. *Cardicola opisthorchis* n. sp. (Trematoda: Aporocotylidae) from the Pacific bluefin tuna, *Thunnus orientalis* (Temminck & Schlegel, 1844), cultured in Japan. Parasitol. Int. 60, 307–312.

Oglesby, L.C., 1961. A new cercaria from an annelid. J. Parasitol. 47, 233–236.

Olson, P.D., Tkach, V.V., 2005. Advances and trends in the molecular systematics of the parasitic Platyhelminthes. Adv. Parasitol. 60, 165–243.

Olson, P.D., Cribb, T.H., Tkach, V.V., Bray, R.A., Littlewood, D.T., 2003. Phylogeny and classification of the Digenea (Platyhelminthes:Trematoda). Int. J. Parasitol. 33, 733–755.

Orélis-Ribeiro, R., Ruiz, C.F., Curran, S.S., Bullard, S.A., 2013. Blood flukes (Digenea: Aporocotylidae) of lamniforms: redescription of *Hyperandrotrema cetorhini* from basking shark (*Cetorhinus maximus*) and description of a new congener from shortfin mako shark (*Isurus oxyrinchus*) off Alabama. J. Parasitol. 99, 835–846.

Overstreet, R.M., Køie, M., 1989. *Pearsonellum corventum*, gen. et sp. nov. (Digenea: Sanguinicolidae), in serranid fishes from the Capricornia section of the Great Barrier Reef. Aust. J. Zool. 37, 71–79.

Platt, T.R., 2002. Spirorchiidae. In: Gibson, D.I., Jones, A., Bray, R. (Eds.), Keys to the Trematoda. CABI Publishing, Wallingford, UK, pp. 453–467.

Platt, T.R., Brooks, D.R., 1997. Evolution of the schistosomes (Digenea: Schistosomatoidea): the origin of dioecy and colonization of the venous system. J. Parasitol. 83, 1035–1044.

Platt, T.R., Sharma, S.K., 2012. Two new species of Hapalorhynchus (Digenea: Spirorchiidae) from Freshwater Turtles (Testudines: Geomydidae) in Malaysia. Comp. Parasitol. 79, 202–207.

Platt, T.R., Blair, D., Purdie, J., Melville, L., 1991. *Griphobilharzia amoena* n. gen., n. sp. (Digenea: Schistosomatidae), a parasite of the freshwater crocodile *Crocodylus johnstoni* (Reptilia: Crodocylia) from Australia, with the erection of a new subfamily, Griphobilharzinae. J. Parasitol. 77, 65–68.

Platt, T.R., Hoberg, E.P., Chisholm, L.A., 2013. On the morphology and taxonomy of *Griphobilharzia amoena* Platt and Blair, 1991 (Schistosomatoidea), a dioecious digenetic trematode parasite of the freshwater crocodile, *Crocodylus johnstoni*, in Australia. J. Parasitol. 99, 888–891.

Polinski, M.P., Hamilton, D.B., Nowak, B.F., Bridle, A., 2013. SYBR, TaqMan, or both: highly sensitive, non-invasive detection of *Cardicola* blood fluke species in Southern Bluefin Tuna (*Thunnus maccoyii*). Mol. Biochem. Parasitol. 191, 7–15.

Polinski, M.P., Shirakashi, S., Bridle, A., Nowak, B.F., 2014. Transcriptional immune response of cage-cultured Pacific Bluefin tuna during infection by two *Cardicola* blood fluke species. Fish Shellfish Immunol. 36, 61–67.

Protasio, A.V., Tsai, I.J., Babbage, A., Nichol, S., Hunt, M., Aslett, M.A., De Silva, N., Velarde, G.S., Anderson, T.J.C., Clark, R.C., Davidson, C., Dillon, G.P., Holroyd, N.E., Loverde, P.T., Lloyd, C., Mcquillan, J., Oliveira, G., Otto, T.D., Parker-Manuel, S.J., Quail, M.A., Wilson, R.A., Zerlotini, A., Dunne, D.W., Berriman, M., 2012. A systematically improved high quality genome and transcriptome of the human blood fluke *Schistosoma mansoni*. PLoS Negl. Trop. Dis. 6, e1455.

Rambaut, A., Drummond, A.J., 2009. Tracer version 1.5. Program. http://beast.bio.ed.ac.uk/Tracer/.

Repullés-Albelda, A., Montero, F.E., Holzer, A.S., Ogawa, K., Hutson, K.S., Raga, J.A., 2008. Speciation of the *Paradeontacylix* spp. (Sanguinicolidae) of *Seriola dumerili*. Two new species of the genus *Paradeontacylix* from Mediterranean. Parasitol. Int. 57, 405–414.

Rollinson, D., Knopp, S., Levitz, S., Stothard, J.R., Tcheuenté, L.T., Garba, A., Mohammed, K.A., Schur, N., Person, B., Colley, D.G., Utzinger, J., 2013. Time to set the agenda for schistosomiasis elimination. Acta Trop. 128, 423–440.

Ronquist, F., Huelsenbeck, J.P., 2003. MRBAYES 3: Bayesian phylogenetic inference under mixed models. Bioinformatics 19, 1572–1574.

Schell, S.C., 1974. The life history of *Sanguinicola idahoensis* sp. n. (Trematoda: Sanguinicolidae), a blood parasite of steelhead trout, *Salmo gairdneri* Richardson. J. Parasitol. 60, 561–566.

Shirakashi, S., Kishimoto, Y., Kinami, R., Katano, H., Ishimaru, K., Murata, O., Itoh, N., Ogawa, K., 2012. Morphology and distribution of blood fluke eggs and associated pathology in the gills of cultured Pacific bluefin tuna, *Thunnus orientalis*. Parasitol. Int. 61, 242–249.

Shirakashi, S., Tsunemoto, K., Rough, K., Webber, C., Ellis, D., Ogawa, K., 2013. Two species of *Cardicola* (Trematoda:Aporocotylidae) found in Southern Bluefin Tuna *Thunnus maccoyi* ranched in South Australia. Fish Pathol. 48, 1–4.

Short, R.B., 1954. A new blood fluke, *Selachohemecus olsoni*, n. g., n. sp. (Aporocotylidae) from the sharp-nosed shark, *Scoliodon terra-novae*. Proc. Helminthol. Soc. Wash. 21, 78–82.

Smith, J.W., 1972. The blood flukes (Digenea: Sanguinicolidae and Spirorchidae) of cold-blooded vertebrates and some comparison with the schistosomes. Helminth. Abstr. 41, 161–204.

Smith, J.W., 1997a. The blood flukes (Digenea: Sanguinicolidae and Spirorchidae) of cold-blooded vertebrates: Part 1. A review of the published literature since 1971, and bibliography. Helminth. Abstr. 66, 255–294.

Smith, J.W., 1997b. The blood flukes (Digenea: Sanguinicolidae and Spirorchidae) of cold-blooded vertebrates: Part 2. Appendix I: comprehensive parasite–host list; Appendix II: comprehensive host–parasite list. Helminth. Abstr. 66, 329–344.

Smith, J.W., 2002. Family Sanguinicolidae von Graff, 1907. In: Gibson, D.I., Jones, A., Bray, R. (Eds.), Keys to the Trematoda. CABI Publishing, Wallingford, UK, pp. 433–452.

Snyder, S.D., 2004. Phylogeny and paraphyly among tetrapod blood flukes (Digenea: Schistosomatidae and Spirorchiidae). Int. J. Parasitol. 34, 1385–1392.

Snyder, S.D., Loker, E.S., 2000. Evolutionary relationships among the Schistosomatidae (Platyhelminthes: Digenea) and an Asian origin for *Schistosoma*. J. Parasitol. 86, 283–288.

Stacy, B.A., Frankovich, T., Greiner, E., Alleman, A.R., Herbst, L.H., Klein, P., Bolten, A., McIntosh, A., Jacobson, E.R., 2010. Detection of spirorchiid trematodes in gastropod tissues by polymerase chain reaction: preliminary identification of an intermediate host of *Learedius learedi*. J. Parasitol. 96, 752–757.

Sugihara, Y., Yamada, T., Tamaki, A., Yamanishi, R., Kanai, K., 2014. Larval stages of the bluefin tuna blood fluke *Cardicola opisthorchis* (Trematoda: Aporocotylidae) found from *Terebella* sp. (Polychaeta: Terebellidae). Parasitol. Int. 6, 295–299.

Tamura, K., Peterson, D., Peterson, N., Stecher, G., Nei, M., Kumar, S., 2011. MEGA5: molecular evolutionary genetics analysis using maximum likelihood, evolutionary distance, and maximum parsimony methods. Mol. Biol. Evol. 28, 2731–2739.

The Schistosoma japonicum Genome Sequencing and Functional Analysis Consortium, 2009. The *Schistosoma japonicum* genome reveals features of host parasite interplay. Nature 460, 345–351.

Tkach, V., Grabda-Kazubska, B., Pawlowski, J., Swiderski, Z., 1999. Molecular and morphological evidence for close phylogenetic affinities of the genera Macrodera, Leptophallus, Metaleptophallus and Paralepoderma (Digenea, Plagiorchiata). Acta Parasitol. 44, 170–179.

Tkach, V.V., Snyder, S.D., Vaughan, J.A., 2009. A new species of blood fluke (Digenea: Spirorchiidae) from the Malayan box turtle, *Cuora amboinensis* (Cryptodira: Geomydidae) in Thailand. J. Parasitol. 95, 743–746.

Truong, T.N., Bullard, S.A., 2013. Blood flukes (Digenea: Aporocotylidae) of walking catfishes (Siluriformes: Clariidae): new genus and species from the Mekong River (Vietnam) and a note on catfish aporocotylids. Folia Parasitol. 60, 237–247.

Tsai, I.J., Zarowiecki, M., Holroyd, N., Garciarrubio, A., Sanchez-Flores, A., Brooks, K.L., Tracey, A., Bobes, R.J., Fragoso, G., Sciutto, E., Aslett, M., Beasley, H., Cai, X., Camicia, F., Clark, R., Cucher, M., De Silva, N., Day, T.A., Deplazes, P., Estrada, K., Fernández, C., Holland, P.W.H., Hou, J., Hu, S., Huckvale, T., Hung, S.S., Kamenetzky, L., Keane, J.A., Kiss, F., Koziol, U., Lambert, O., Liu, K., Luo, X., Luo, Y., Macchiaroli, N., Nichol, S., Paps, J., Parkinson, Jn., Pouchkina-Stantcheva, N., Riddiford, N., Rosenzvit, M., Salinas, G., Wasmuth, J.D., Zamanian, M., Zheng, Y., The Taenia solium Genome Consortium, Cai, J., Soberón, X., Olson, P.D., Laclette, J.P., Brehm, K., Berriman, M., 2013. The genomes of four tapeworm species reveal adaptations to parasitism. Nature 496, 57–63.

Van der Land, J., 1967. A new blood fluke (Trematoda) from *Chimaera monstrosa* L. Proc. Koninklijke Akademie van Wetenschappen te Amsterdam, Section C. Biol. Med. Sci. 70, 110–120.

Van Der Werf, M.J., De Vias, S.J., Brooker, S., Looman, C.W., Nagelkerke, N.J., Habbema, J.D., Engels, D., 2003. Quantification of clinical morbidity associated with schistosome infection in sub-Saharan Africa. Acta Trop. 86, 125–139.

Wales, J.H., 1958. Two new blood flukes of trout. Calif. Fish Game 44, 125–136.

Wang, C.R., Li, L., Ni, H.B., Zhai, Y.Q., Chen, A.H., Chen, J., Zhu, X.Q., 2009. *Orientobilharzia turkestanicum* is a member of Schistosoma genus based on phylogenetic analysis using ribosomal DNA sequences. Exp. Parasitol. 121, 193–197.

Wardle, W.J., 1979. A new marine cercaria (Digenea: Aporocotylidae) from the southern quahog *Mercenaria campechiensis*. Contrib. Mar. Sci. 22, 53–56.

Webster, B.L., Littlewood, D.T.J., 2012. Mitochondrial gene order change in Schistosoma (Platyhelminthes: Digenea: Schistosomatidae). Int. J. Parasitol. 42, 313–321.

Webster, B.L., Southgate, V.R., Littlewood, D.T.J., 2006. A revision of the interrelationships of Schistosoma including the recently described *Schistosoma guineensis*. Int. J. Parasitol. 36, 947–955.

World Health Organization, 2010. Working to Overcome the Global Impact of Neglected Tropical Diseases—First WHO Report on Neglected Tropical Diseases. World Health Organization, Geneva.

Yong, R.Q.Y., Cribb, T.H., 2011. *Rhaphidotrema kiatkiongi*, a new genus and species of blood fluke (Digenea: Aporocotylidae) from *Arothron hispidus* (Osteichthyes: Tetraodontidae) from the Great Barrier Reef, Australia. Folia Parasitol. 58, 273–277.

Yong, R.Q., Cutmore, S.C., Miller, T.L., Adlard, R.D., Cribb, T.H., 2013. The ghost of parasites past: eggs of the blood fluke *Cardicola chaetodontis* (Aporocotylidae) trapped in the heart and gills of butterflyfishes (Perciformes: Chaetodontidae) of the Great Barrier Reef. Parasitology 140, 1186–1194.

Young, N.D., Jex, A.R., Li, B., Liu, S., Yang, L., Xiong, Z., Li, Y., Cantacessi, C., Hall, R.S., Xu, X., Chen, F., Wu, X., Zerlotini, A., Oliveira, G., Hofmann, A., Zhang, G., Fang, X., Kang, Y., Campbell, B.E., Loukas, A., Ranganathan, S., Rollinson, D., Rinaldi, G., Brindley, P.J., Yang, H., Wang, J., Wang, J., Gasser, R.B., 2012. Whole-genome sequence of *Schistosoma haematobium*. Nat. Genet. 44, 221–225.

CHAPTER TWO

Techniques for the Diagnosis of *Fasciola* Infections in Animals: Room for Improvement

Cristian A. Alvarez Rojas*, Aaron R. Jex†, Robin B. Gasser†,[1], Jean-Pierre Y. Scheerlinck*

*Centre for Animal Biotechnology, Faculty of Veterinary Science, The University of Melbourne, Parkville, Victoria, Australia

†Faculty of Veterinary Science, The University of Melbourne, Parkville, Victoria, Australia

[1]Corresponding author: e-mail address: robinbg@unimelb.edu.au

Contents

Abstract

The common liver fluke, *Fasciola hepatica*, causes fascioliasis, a significant disease in mammals, including livestock, wildlife and humans, with a major socioeconomic impact worldwide. In spite of its impact, and some advances towards the development of

Advances in Parasitology, Volume 85
ISSN 0065-308X
http://dx.doi.org/10.1016/B978-0-12-800182-0.00002-7

vaccines and new therapeutic agents, limited attention has been paid to the need for practical and reliable methods for the diagnosis of infection or disease. Accurate diagnosis is central to effective control, particularly given an emerging problem with drug resistance in *F. hepatica*. Traditional coprological techniques have been widely used, but are often unreliable. Although there have been some advances in establishing immunologic techniques, these tools can suffer from a lack of diagnostic specificity and/or sensitivity. Nonetheless, antigen detection tests seem to have considerable potential, but have not yet been adequately evaluated in the field. Moreover, advanced nucleic acid-based methods appear to offer the most promise for the diagnosis of current infection. This chapter (i) provides a brief account of the biology and significance of *F. hepatica*/fascioliasis, (ii) describes key techniques currently in use, (iii) compares their advantages/disadvantages and (iv) reviews polymerase chain reaction-based methods for specific diagnosis and/or the genetic characterization of *Fasciola* species.

1. INTRODUCTION

Parasites of animals often cause diseases of major socioeconomic importance worldwide. These diseases have a substantial adverse impact on farm profitability and can jeopardize sustainable food production. For instance, the annual cost associated with *Fasciola* species, one of the parasitic trematodes, had been estimated at ~\$3 billion dollars worldwide (FAO, 1994). Thus, there are substantial economic gains to be made in livestock production by enhancing the control of *Fasciola hepatica* and related parasites. Currently, these trematodes are mainly controlled by chemotherapeutic agents (anthelmintics) such as closantel (Maes et al., 1988) and triclabendazole (Boray et al., 1983). Even with optimally timed (strategic) treatments, this approach is expensive for farmers and only partially effective. In addition, the excessive use of triclabendazole can result in drug resistance (Alvarez-Sanchez et al., 2006; Brockwell et al., 2014; Moll et al., 2000; Overend and Bowen, 1995). Therefore, there is a clear need, and considerable global interest, in the development of improved methods for the integrated control of fascioliasis, which should incorporate the use of specific and sensitive diagnostic tools.

The accurate diagnosis of *Fasciola* infections is central to studying the epidemiology of fascioliasis and the surveillance and control of this disease. The classical parasitological methods of diagnosis, such as the microscopic detection of *Fasciola* eggs, are widely and routinely used (Martinez-Perez et al., 2012), but are relatively unreliable and time-consuming to perform. On the other hand, methods for the detection of anti-*Fasciola* serum antibody

(Ambrorse-Thomas et al., 1980; Chauvin et al., 1997) are suitable for investigations of prevalence at a herd level. Methods involving the specific detection of *Fasciola* antigens in serum or faeces can detect current infections, and DNA-based technologies appear to provide avenues to overcome current limitations and to develop substantially improved tools for the diagnosis of *F. hepatica* infections, utilizing specific genetic markers (Ai et al., 2010a; Le et al., 2012a; Robles-Perez et al., 2013). In this chapter, we (i) provide a brief account of the biology and significance of *F. hepatica*/fascioliasis, (ii) describe some key techniques in current use for the diagnosis of infection/disease in the definitive host, (iii) compare their advantages and disadvantages and (iv) review the developments in polymerase chain reaction (PCR)-based methods for specific diagnosis and the genetic characterization of *Fasciola* species.

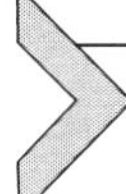

2. A BRIEF BACKGROUND ON FASCIOLIASIS AND THE BIOLOGY OF *Fasciola* SPECIES

Liver flukes are important parasitic flatworms (Platyhelminthes: Trematoda) affecting animals in a wide range of countries. *Fasciola* spp. are known to infect a wide variety of mammals (definitive hosts), including ruminants, suidians, primates, elephants, hippopotami, lagomorphs and rodents (Mas-Coma et al., 2009; Menard et al., 2000), some being more permissive than others. The two representatives of most key socioeconomic importance are *F. hepatica* and *Fasciola gigantica*. These two species are usually the causative agents of fascioliasis in livestock, wildlife and humans (Mas-Coma et al., 2009; Torgerson and Claxton, 1999). Fascioliasis is recognized as a neglected disease and occurs mainly in parts of Africa, the Middle East, South America and Southeast Asia (Mas-Coma, 2005; Mas-Coma et al., 2009).

F. hepatica and *F. gigantica* are similar morphologically and biologically (Itagaki et al., 2009; Le et al., 2008; Peng et al., 2009). Both *Fasciola* spp. have di-heteroxenous life cycles, which involve (freshwater) lymnaeid snails as intermediate hosts and mammals as definitive hosts (Andrews, 1999). Differences in host specificity between *F. gigantica* and *F. hepatica* appear to define the aetiology and clinical manifestation of disease in the definitive host (Spithill et al., 1999). A key difference between these parasites is their adaptation to different intermediate snail hosts, linked to the geographic distribution of these parasites and disease. *F. hepatica* often utilizes snails such as *Lymnaea tomentosa* and *Galba truncatula*, which are widespread in temperate and subtropical climes (Mas-Coma et al., 2009). *F. gigantica* usually prefers snails, including *Radix rubiginosa* and *R. natalensis*, which live in the subtropics and tropics. In subtropical regions, both species of *Fasciola* can coexist, and fascioliasis can also be associated with *Fasciola* sp. (a proposed *F. gigantica* × *F. hepatica* hybrid) (cf. Mas-Coma et al., 2009; Spithill et al., 1999).

3. PATHOGENESIS AND CLINICAL DIAGNOSIS OF FASCIOLIASIS

The pathogenesis of fascioliasis in the mammalian definitive host is usually divided into acute and chronic phases. The *acute/subacute phase* (1–6 weeks post infection, WPI) commences with the ingestion of the metacercarial (larval) stage on herbage or in water and is characterized by the migration of immature worms through the duodenal wall into the abdominal cavity and then through the liver capsule and parenchyma, resulting in tissue damage and leading to traumatic hepatitis (Boray, 1969). Sheep are more affected by the acute phase of infection than cattle. The liver enzymes glutamate dehydrogenase and gamma-glutamyl transpeptidase (GGT) are released into the circulation due to the destruction of the liver during the migration of the parasite (Anderson et al., 1977; Sykes et al., 1980; Thorpe and Ford, 1969) and can be measured to support clinical diagnosis (Lotfollahzadeh et al., 2008). Clinical signs can include abdominal pain, fever, anaemia, hepatomegaly, diarrhoea, reduced weight gain and/or ill thrift; however, none of these clinical signs is pathognomonic for fascioliasis. Indeed, in ruminants, similar manifestations can be caused by other parasites, including nematodes such as *Haemonchus*, *Teladorsagia*, *Ostertagia*, *Cooperia* and *Trichostrongylus* (Hungerford, 1990).

The *chronic phase* commences when adult worms have established in the biliary ducts, occurring typically from 7 to 8 WPI (Boray, 1969). Elevated levels of GGT are an indicator of epithelial damage in the bile ducts by *Fasciola* (Elliott et al., 2013). In addition to hepatic fibrosis (following acute or subacute infection) and anaemia, the chronic phase is characterized by progressive cholangitis, hyperplasia of the duct epithelium and periductal fibrosis, cholestasis and/or cholelithiasis (Behm and Sangster, 1999). Fascioliasis can also sometimes be associated with complications, such as co-infections with bacteria (e.g. *Clostridium novyi*, causing "black disease" in sheep) (Boray, 1969; Carpenter, 1998).

The clinical manifestation of fascioliasis in definitive hosts depends on parasite factors (e.g. species/strain of worm, infective dose and/or intensity of infection) and host factors (e.g. species of host, genetic composition, immune response and phase/duration of the infection) (Haroun and Hillyer, 1986; Piedrafita et al., 2004). Most breeds of sheep are highly susceptible to fascioliasis caused by *F. hepatica* (Haroun and Hillyer, 1986), and studies suggest that *F. gigantica* may be better adapted to cattle, with higher levels of resistance being observed in some breeds of sheep and goats (Haroun et al., 1989; Piedrafita et al., 2004; Roberts et al., 1997). For instance, the Indonesian thin tail sheep shows an acquired or innate resistance against *F. gigantica* infection and develops a high level of resistance after a primary challenge (Roberts et al., 1997).

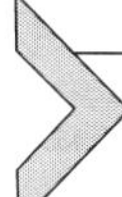

4. CLASSICAL METHODS FOR DETECTION OF *F. hepatica* INFECTION

Ante-mortem diagnosis of *F. hepatica* infection in individual animals is usually based on the microscopic detection or enumeration of parasite eggs in the faeces following concentration by sedimentation or combined flotation/sedimentation (Happich and Boray, 1969a,b; McCaughey and Hatch, 1964; Sewell and Hammond, 1972). While such methods are often established in routine diagnostic laboratories, commercial tests are available and include FLOTAC (Cringoli et al., 2010) and Flukefinder (Foreyt, 2002). In spite of their utility, it needs to be emphasized that eggs are not detected during a pre-patent infection (with juvenile worms). Eggs are detected during patent infection (with mature and gravid worms) but can be released sporadically from the parasite in the bile ducts and worms can differ in their biotic potential/fecundity, depending on their stage of development, the host that they infect and/or immune attack by the host (Gonzalez-Lanza et al., 1989; Sewell and Hammond, 1972). Moreover, in Asia, offspring originating from hybridization between *F. hepatica* and *F. gigantica* have been shown to be aspermic (Peng et al., 2009) and are thus expected to be infertile, although some experiments have showed fertile hybrids (producing less eggs than the parental parasites) (Kuboki et al., 2003). In addition, other factors, including host age (Honer, 1965c), faecal water content (Honer, 1967) and the number of aliquots tested per sample (Honer, 1965b), can contribute to the variability in detecting *Fasciola* eggs in faeces, with more samples tested increasing the sensitivity of detection (Rapsch et al., 2006). All of these factors can result in recurring false-negative results, which compromise the sensitivity of this classical diagnostic approach (Dorsman, 1960; Honer, 1965a; Mas-Coma et al., 1999). Hence, taken together, these issues indicate that the number of *Fasciola* eggs per gram of faeces does not necessarily relate to the number of worms present in the liver parenchyma or bile ducts of an infected host.

Further complicating the specific diagnosis of *Fasciola* infection by coprological examination, the similarity in structure, size and shape of the *Fasciola* eggs prevents their specific identification (on an individual basis) in areas where *F. hepatica* and *F. gigantica* co-occur (Adela Valero et al., 2009). On rare occasions, false test-positive results might occur as a consequence of pseudo-parasitism (i.e. following the consumption of liver from an infected animal) (Hillyer, 1998).

Another key method for assessing *Fasciola* infection intensity is the *post-mortem* examination of animals (Urquhart et al., 1996) that have succumbed or been selected from a flock/herd and euthanized. Livers are examined for juvenile and bile ducts for adult worms and associated pathological changes. For the acute phase of infection/disease, these alterations include fibrinous exudate on the liver capsule, bleeding, tissue destruction and inflammation (Marcet Sanchez et al., 2012). Later on, during the infection, more evidence of tissue repair or atrophy can be detected, together with swelling/collapse in some liver lobes and fibrin tags (Boray, 1969). This extends to the chronic infection, where the parasite has resided within the host for more than 4–6 months, and adults can be found in the major bile ducts, leading to prolonged inflammation, resulting in biliary hyperplasia and periductal fibrosis. In high-intensity infections, occlusion of the bile ducts, reduced bile flow/drainage and/or significant enlargement of the gall bladder can occur (Njiru et al., 2008).

5. IMMUNODIAGNOSIS

The assessment of the immune response in live animals has been widely used to infer current infection or previous exposure to the parasite. Early attempts to diagnose fascioliasis in animals commenced in the beginning of the twentieth century, with the assessment of skin reactions following the injection of somatic extracts of *F. hepatica* into the host (Curasson, 1935; Sievers and Oyarzun, 1932). However, this approach was shown to be unreliable, as cross-reactivity with other parasites is common (Aygun and Baskaya, 1939; Curasson, 1935; Lavier and Stefanopoulo, 1944; Sievers and Oyarzun, 1932; Soulsby, 1954; Szaflarski, 1950; Wagner, 1935). Subsequently, the detection of specific antibodies against parasite antigens in the serum of infected individuals became more common.

Techniques included, for example, complement fixation (Benex et al., 1959; Lavier and Stefanopoulo, 1944), immunoelectrophoresis (Teodorovic et al., 1963), precipitin reaction (Korach and Benex, 1966), immunofluorescence (Coudert et al., 1967; Deelder and Ploem, 1975), double immunodiffusion (Tailliez, 1967), haemagglutination (Hillyer and Allain, 1979; Levieux et al., 1992b), enzyme-linked immunosorbent assay (ELISA) (Arias et al., 2006; Cornelissen et al., 1999, 2001; Espino and Hillyer, 2003; Gaudier et al., 2012; Hillyer and Serrano, 1986; Kuerpick et al., 2013; Martinez et al., 1996; Martinez-Moreno et al., 1997; Mezo et al., 2007; Paz-Silva et al., 2005; Reichel, 2002; Ruiz et al., 2003; Salimi-Bejestani et al., 2007; Santiago and Hillyer, 1988; Sexton et al., 1991; Silva et al., 2004; Timoteo et al., 2005; Zimmerman et al., 1982) and immunoblotting (Bossaert et al., 2000; Espino and Hillyer, 2003; Fredes et al., 1997; Gorman et al., 1997; Hammami et al., 1997; Ortiz et al., 2000; Qureshi et al., 1995; Sampaio-Silva et al., 1996; Santiago and Hillyer, 1988).

5.1. Antigens used in immunological assays

A wide range of antigens, including excretory/secretory (ES) products, tegumental components, crude extracts from adult worms and, more recently, recombinant proteins, have been used in various assays (Table 2.1).

5.1.1 ES antigens

ES products from *F. hepatica* have been widely used as antigens in serological assays. These products represent a complex mixture of molecules that have been actively transported via secretory pathways, or diffuse from the parasite soma to the exterior of the parasite, and those that slough from layers of the tegument and/or sub-tegument (Mulcahy et al., 1999). The antigenic complexity of ES products was recognized by Lehner and Sewell (1980), who used them to assess anti-*Fasciola* serum antibody responses in rabbits, rats and sheep. Other studies (Cervi et al., 1992; de Weil et al., 1984; Zimmerman and Clark, 1986) suggested that fractionations of ES products or single antigenic components would achieve improved diagnostic performance. Gundlach (1971), for instance, reported that ES antigens gave better serological test results than somatic antigens for the diagnosis of pre-patent *F. hepatica* infection in rabbits. In addition, Sandeman and Howell (1981) identified antigens in ES products that were shared between juvenile and adult stages of *F. hepatica* for the detection of antibodies in the sera from infected sheep. Some major antigenic components of ES products included a leucine aminopeptidase and a phosphoenolpyruvate carboxykinase

Table 2.1 Antigens from *Fasciola hepatica* recognized by antibodies in sera from different infected animals

Antigen	Molecular weight (kDa)/antigen name	Animal sera	References
ES (excretory/ secretory) products	23, 24 and 26	Sheep	Irving and Howell (1982)
	25 and 30	Rabbit, sheep, cattle	Santiago et al. (1986)
	29	Sheep, pig, horse	Fredes et al. (1997)
	14, 17, 22, 30, 40 and 42	Horse, pigs	Gorman et al. (1997)
	25–30	Rabbit	Rivera Marrero et al. (1988)
	20–28	Sheep, cattle	Santiago and Hillyer (1988)
	150–160	Cattle, sheep	Santiago and Hillyer (1988)
	20–28	Cattle, sheep	Santiago and Hillyer (1988)
	17	Human, rabbit, cattle, sheep	Hillyer and Soler de Galanes (1988)
	29–31	Sheep	Sexton et al. (1991)
		Human	Hillyer et al. (1992)
	11.5, 46 and 67	Rabbit	Espino and Hillyer (2003)
	15	Cattle	Qureshi et al. (1995)
	25 and 27	Human	Sampaio-Silva et al. (1996)
	29 and 57	Human	Hammami et al. (1997)
	26–30	Cattle	Bossaert et al. (2000)
	28	Cattle	Ortiz et al. (2000)
	66 and 17	Cattle	Ortiz et al. (2000)

Table 2.1 Antigens from *Fasciola hepatica* recognized by antibodies in sera from different infected animals—cont'd

Antigen	Molecular weight (kDa)/antigen name	Animal sera	References
Cysteine proteinases	Fas1 (26), Fas2 (25)	Human	Cordova et al. (1997, 1999) and Espinoza et al. (2007, 2005)
	Fas1 (26), Fas2 (25)	Alpacas	Neyra et al. (2002) and Timoteo et al. (2005)
	Cathepsin L1	Human	O'Neill et al. (1998) and Rokni et al. (2002)
Tegumental antigens	Enolase, aldolase, glutathione *S*-transferase, fatty acid-binding protein	Rabbits, human	Hillyer (1980), Hillyer and Serrano (1986) and Morales and Espino (2012).
	T1	Sheep, cattle	Trudgett et al. (1988)
	FhTP16.5		Gaudier et al. (2012)
Crude antigens	30–38, 56, 64 and 69	Cattle, sheep	Santiago and Hillyer (1988)
	N/A	Rabbits	Hillyer and Cervoni (1978)
	14–19, 22–30, 35–37 and 42	Horse, pig	Gorman et al. (1997)
	8, 9, 10, 38 and 45 from total soluble extract (FhTSE) and 8, 12, 15 and 24 from the adult worm vomit (FhAWV)	Human	De Almeida et al. (2007)
	8	Human	Kim et al. (2003)
	N/A	Human	Khalil et al. (1990)
	Fasciola worm antigen preparation (FWAP)	Human	Maher et al. (1999)
Other and recombinant antigens	Metacercarial antigens	Mice	Bennett et al. (1982)
	Ferritin (FhFtn-1)	Rabbits	Caban-Hernandez et al. (2012)

Continued

Table 2.1 Antigens from *Fasciola hepatica* recognized by antibodies in sera from different infected animals—cont'd

Antigen	Molecular weight (kDa)/antigen name	Animal sera	References
	Heat shock protein (Fh-HSP35a) 65	Cattle, goat, rat	Moxon et al. (2010)
	Cystatin	Human	Ikeda (1998)
	Rec-cathepsin L1 (SeroFluke)	Human	Martinez-Sernandez et al. (2011)
	rFh8 8	Rabbit, rat, cattle, sheep and human	Silva et al. (2004)
	rFhSAP2 11.5	Rabbit	Espino and Hillyer (2003)
	rAPS	Sheep, horse	Arias et al. (2006, 2012)

(Marcilla et al., 2008). The main issue when using ES products is batch-to-batch variation in the quality and amount of antigens. Espino et al. (1987) suggested that antigens should be collected after 24 h of culture, in order to decrease cross-reactivity in immunologic assays.

5.1.2 Cysteine proteases

These peptidases are major components of the proteome of *Fasciola* spp. and play central roles in the biology of these parasites (Robinson et al., 2008), such as migratory and digestive processes (Collins et al., 2004; Yamasaki et al., 1989). These proteins can also be found consistently in ES products (Berasain et al., 1997, 2000; Carmona et al., 1993; Corvo et al., 2009; Piacenza et al., 1997). Numerous studies have focused on the isolation of specific cysteine proteinases and on assessing their potential value as diagnostic antigens. For example, two purified ES cysteine proteinases, Fas1 (26 kDa) and Fas2 (25 kDa) (Cordova et al., 1997), were used in ELISA for the diagnosis of fascioliasis (Cordova et al., 1999; Espinoza et al., 2005, 2007) and in alpacas (Neyra et al., 2002; Timoteo et al., 2005). These proteins are also present in the arc2 precipitation band diagnostic in some conventional tests (Muro et al., 1994; Santiago de Weil et al., 1984). Cathepsin L1 has also been used as an antigen for diagnosis in Bolivia (O'Neill et al., 1998) and Iran (Rokni et al., 2002). The expression of cathepsin L1 in various helminths, such as cestodes, suggests that serological cross-reactivity could be an issue (Cornelissen et al., 1999).

5.1.3 Tegumental antigens

Molecules in, or released as ES products from, the tegument of *F. hepatica* are in intimate contact with the immune system of the host (Lammas and Duffus, 1983). Hillyer (1980) first isolated a tegumental antigen of *F. hepatica* (Fh_t) and used it in immunodiffusion tests, detecting positive serological reactions in infected rabbits, but cross-reactivity occurred (see Hillyer and Serrano, 1986). Major components of this antigen are enolase, aldolase, glutathione *S*-transferase and fatty acid-binding protein (Morales and Espino, 2012). Another tegumental antigen (designated T1) was isolated and used in ELISA and allowed the detection of serum antibodies in sheep and cattle (Trudgett et al., 1988). More recently, a tegumental protein, called FhTP16.5, was characterized and localized and also found in ES products of *F. hepatica* (see Gaudier et al., 2012).

5.1.4 Somatic antigens

Antigens from crude extracts from the soma of adult worms have been used in immunologic assays for the detection of serum antibodies in cattle, sheep or rabbits between 2 and 4 WPI (Hillyer and Cervoni, 1978; Hillyer and Taylor, 1988) (Table 2.2). Gorman et al. (1997) characterized antigenic components of 14–19, 22–30, 35–37 and 42 kDa in somatic antigens using sera from *F. hepatica*-infected horses and pigs. In contrast, antigenic fractions of 8, 9, 10, 38 and 45 kDa from total soluble extract (FhTSE) and 8, 12, 15 and 24 kDa from the adult worm "vomit" (FhAWV) were found to be recognized by sera from *F. hepatica*-infected patients (De Almeida et al., 2007). Although relatively specific (Kim et al., 2003; Maher et al., 1999), cross-reactivity with other trematodes was suggested for the 8 kDa fraction. Cross-reactivity has also been reported for other somatic antigens (Khalil et al., 1990).

5.1.5 Other parasite antigens

Various other antigens that have been used for diagnostic purposes include (i) metacercariae antigens (Bennett et al., 1982); (ii) a ferritin from *F. hepatica* (FhFtn-1), with strong reactivity with sera from infected rabbits with acute and chronic infections (Caban-Hernandez et al., 2012); (iii) a heat shock protein (Fh-HSP35a), suggested to be used as an egg-based diagnostic

Table 2.2 Summary of detection times of specific antibodies against different antigens of *Fasciola hepatica* in different animal species experimentally infected

Species	Antigens	Antibody detection time	References
Sheep	ES	2 WPI	Zimmerman et al. (1982)
Sheep	ES and somatic	2–20 WPI	Cornelissen et al. (1992)
Sheep	ES	2–20 WPI	Sexton et al. (1991)
Cattle	ES	2–10 WPI	Hillyer and Soler de Galanes (1991)
Sheep	ES	2–10 WPI	Hillyer and Soler de Galanes (1991)
Rabbit	ES	3–16 WPI	Hillyer and Soler de Galanes (1991)
Mice	ES	2–5 WPI	Hillyer and Soler de Galanes (1991)
Goat	ES	2–4 WPI to 1 year	Martinez et al. (1996)
Goat	ES	2–20 WPI	Martinez-Moreno et al. (1997)
Cattle	ES	4 WPI	Farrell et al. (1981)
Rabbit	ES (Fhm2)	12–103 DPI	Pfister et al. (1984)
Cattle	ES and somatic (Fhwwe)	2–10 WPI	Santiago and Hillyer (1988)
Sheep	ES and somatic (Fhwwe)	4–10 WPI	Santiago and Hillyer (1988)
Rabbit	rFh8 (ES)	3–14 WPI	Silva et al. (2004)
Sheep	rFh8 (ES)	5–7 WPI	Silva et al. (2004)
Rabbit	Tegument (Fht)	4–20 WPI	Hillyer (1980)
Rabbit	Tegument (Fht)	2–18 WPI	Hillyer and Serrano (1986)
Rabbit	Tegument (FhTP16.5)	3–12 WPI	Gaudier et al. (2012)
Sheep	Tegument (FhrAPS)	2–30 WPI	Paz-Silva et al. (2005)
Sheep	Tegument (FhrAPS)	3–34 WPI	Arias et al. (2006)
Rabbit	Somatic (Fhs1)	19–103 DPI	Pfister et al. (1984)
Rabbit	FhsmIII	2–21 WPI	Hillyer and Taylor (1988)
Rabbit	rFhSAP	4 WPI	Espino and Hillyer (2003)

Table 2.2 Summary of detection times of specific antibodies against different antigens of *Fasciola hepatica* in different animal species experimentally infected—cont'd

Species	Antigens	Antibody detection time	References
Cattle	F2	2–28 WPI	Levieux et al (1992a)
Goats	F2	2–15 WPI	Levieux and Levieux (1994)
Cattle	F2 (Pourquier)	2–12 WPI	Reichel (2002)
Cattle	F2 (Pourquier)	4–21 WPI	Kuerpick et al. (2013)
Llama	Somatic	2 WPI	Rickard (1995)
Alpaca	Fas1, Fas2	2–24 WPI	Timoteo et al. (2005)
Cattle	Peptides CL1, ES, Purified CL1	3–26 WPI	Cornelissen et al. (1999)
Cattle	rCL1	5–23 WPI	Cornelissen et al. (2001)
Goat	28-kDa cysteine proteinase	2–19 WPI	Ruiz et al. (2003)
Goat	34-kDa cysteine proteinase	4–18 WPI	Ruiz et al. (2003)
Sheep	MM3-SERO	3–18 WPI	Mezo et al. (2007)

ES, excretory/secretory; WPI, week(s) post infection; DPI, days post infection.

marker (Moxon et al., 2010); and (iv) cystatin, a reversible competitive and tight binding inhibitor of cysteine proteases (Ikeda, 1998).

5.1.6 Recombinant antigens

Such antigens have been used to attempt to overcome the disadvantage of variation in quality and quantity of antigens derived directly from *Fasciola*. Recombinant proteins produced for diagnostic purposes have included (i) a recombinant form of an L cathepsin protease (O'Neill et al., 1999); (ii) cathepsin L1 from the adult stage of *F. hepatica* (see Martinez-Sernandez et al., 2011; Muino et al., 2011) to develop a lateral flow immunoassay (SeroFluke); (iii) a recombinant antigen (rFh8) from ES products (Silva et al., 2004); (iv) a saposin-like protein from *F. hepatica* (rFhSAP2) (Espino and Hillyer, 2003); and (v) a recombinant protein of 2.9 kDa expressed from a cDNA library of *F. hepatica* (FhrAPS). Arias et al. (2006) used FhrAPS to detect anti-*Fasciola* IgG antibodies in sheep sera and described better diagnostic sensitivity and specificity in ELISA for FhrAPS as an antigen compared with ES products (Arias et al., 2007), suggesting its suitability for application in seroprevalence studies. Arias et al. (2012) continued this research using the FhrAPS protein, and found a positive correlation between seroprevalence and age of the animals infected in a study of naturally infected horses. An important consideration when using recombinant antigens is that the glycosylation of such proteins can vary considerably, also from native *Fasciola* antigens, depending on the (prokaryotic or eukaryotic) expression system employed to produce them. Such variation has the potential to affect the performance of a diagnostic assay (Huang et al., 2012).

5.2. Detection of specific antibodies against *F. hepatica* in serum

As *F. hepatica* induces a humoral immune response in the infected host, the majority of serological methods for diagnosis of *F. hepatica* infection rely on the detection of antibodies in serum using antigens derived from *F. hepatica*. A disadvantage of this diagnostic approach is the inability to differentiate current from previous infection or exposure (Espino et al., 1997a; Rodriguez-Perez and Hillyer, 1995). Another issue is serological cross-reactivity of anti-*Fasciola* antibodies with antigens from related trematodes, such as species of *Schistosoma* (Hillyer and Serrano, 1983; Hillyer and Soler de Galanes, 1988; Ikeda, 1998), *Paragonimus* (Hillyer and Serrano, 1983), *Paramphistomum* (Ibarra et al., 1998) and *Opisthorchis* (Espino et al., 1997b); the cestodes *Echinococcus granulosus* (see Bossaert et al., 2000; Khalil et al., 1990; O'Neill et al., 1998), *Taenia hydatigena* and *T. ovis* (see Bossaert et al., 2000); and nematodes, such as *Trichinella spiralis* (Hillyer and Soler de Galanes, 1988), *Toxocara canis* (Romasanta et al., 2003) and *Ascaris suum* (Romasanta et al., 2003). Mammalian-type and Gal(β1-6)Gal glycolipids have been inferred as causes of cross-reactivity with cestodes (Wuhrer et al., 2004). In experiments performed by various authors (Table 2.2), anti-*F. hepatica* serum antibodies can be detected from 2 WPI and can remain at high levels after 20 WPI in different definitive hosts of *F. hepatica*. High sensitivity and specificity have been reported in the literature for different antibody-based detection tests (Table 2.3); most results were derived from experimental infections.

5.2.1 Declining serum antibody following anthelmintic treatment

Serum antibody levels can decrease following treatment against fascioliasis or *F. hepatica* infection, although there is considerable variation in findings

Table 2.3 Sensitivity and specificity values reported in the literature using different antigenic preparations for the detection of antibodies against *Fasciola hepatica*

Species	Antigen	Sensitivity	Specificity	References
Sheep	ES	100%	100%	Bautista-Garfias et al. (1989)
Sheep	ES	68.2%	100%	Bautista-Garfias et al. (1989)
Sheep	ES	N/A	95%	Cornelissen et al.(1992)
Sheep	IHA	N/A	86%	Cornelissen et al. (1992)
Human	ES	95%	97%	Sampaio-Silva et al. (1996)
Human	ES 57 kDa	79%	100%	Hammami et al. (1997)
Human	ES 29 kDa	93%	100%	Hammami et al. (1997)
Cattle	ES	92%	N/A	Bossaert et al. (2000)
Human	ES	100%	100%	Carnevale et al. (2001)
Human	ES	100%	97%	Carnevale et al. (2001)
Human	ES	100%	100%	Haseeb et al. (2003)
Human	DRG kit	95.3%	97.5%	Valero et al. (2012b)
Sheep	FhrAPS	92%	86%	Arias et al. (2007)
Sheep	FhrAPS	91%	86%	Arias et al. (2006)
Horse/ donkey	FhrAPS	83%	86%	Arias et al. (2012)
Human	Fas1	89	98	Cordova et al. (1999)
Human	Fas2	95	100	Cordova et al. (1999)
Human	Fas2	95.5	86.6%	Espinoza et al. (2005)
Human	Fas2	92.4	83.6%	Espinoza et al. (2007)
Alpaca	Fas1	90	N/A	Neyra et al. (2002)
Alpaca	Fas2	95	N/A	Neyra et al. (2002)
Human	Cathepsin L	100%	100%	Rokni et al. (2002)
Cattle	CL1 peptide	98.9%	99.8%	Cornelissen et al. (1999)
Cattle	CL1	100%	94.6%	Cornelissen et al. (1999)
Sheep	rCL1	99.1%	98.5%	Cornelissen et al. (2001)
Cattle	rCL1	100%	96.5%	Cornelissen et al. (2001)

Continued

Table 2.3 Sensitivity and specificity values reported in the literature using different antigenic preparations for the detection of antibodies against *Fasciola hepatica*—cont'd

Species	Antigen	Sensitivity	Specificity	References
Human	rCL1	100%	100%	Heussler and Dobbelaere (1994)
Human	CL1	100%	100%	Martinez-Sernandez et al. (2011)
Cattle	MM3	99.2%	100%	Mezo et al. (2010b)
Human	CL1	100%	98%	Yamasaki et al. (1989)
Sheep	F2 (Pourquier)	98%	N/A	Reichel (2002)
Cattle	F2 (Pourquier)	100%	N/A	Reichel (2002)
Cattle (milk)	F2 (Pourquier)	95%	98.2%	Reichel et al. (2005)
Cattle	F2 (Pourquier)	98.2%	98.3%	Molloy et al. (2005)
Cattle (milk)	F2 (Pourquier)	97.7%	99.3%	Molloy et al. (2005)
Sheep	F2 (Pourquier)	96.9%	99.4%	Molloy et al. (2005)
Cattle	F2 (Pourquier)	91.7%	93.7%	Rapsch et al. (2006)
Cattle	F2 (Pourquier)	100%	100%	Kuerpick et al. (2013)

ES, excretory/secretory.

among studies. For instance, some investigations have shown a decrease in antibody levels at 3–6 months after treatment of cattle (Chauvin et al., 1997), 4–6 months after treatment with triclabendazole in cattle (Castro et al., 2000), 1-week post treatment (WPT) with triclabendazole using the recombinant protein FhrAPS in ELISA, 3 WPT using ES products and 4 WPT with triclabendazole in sheep using the MM3 kit (Mezo et al., 2007) and also using the FhrAPS antigen in sheep and cattle treated with albendazole and netobimin (Arias et al., 2009). In other studies of experimentally infected cattle (after 20 WPI), antibody levels decreased, slowing from 4 to 40 WPT with nitroxynil (Boulard et al., 1995). However, in naturally infected cows ($n=45$), no significant decrease in serum antibody (IgG) levels was detected after nitroxynil or closantel treatment (Boulard et al., 1995). Hillyer and Soler de Galanes (1988) reported that antibody levels (IgG, IgM and IgA) against a 17 kDa protein can decrease following treatment. These antibody levels reached normal values between 20 and 47 WPT.

However, contradictory results were described by the same authors using an ES-based ELISA test, with no significant difference found in the antibody levels in patients before and 3 months after treatment (Espino et al., 1992).

5.3. Specific detection of antibodies in milk

Specific antibody has also been detected in milk samples from dairy cows. A negative association between the level of anti-*Fasciola* (IgG) antibodies in bulk milk and the average milk production of the herd was observed (Mezo et al., 2011). This correlation could be used to estimate economic losses suffered from fascioliasis on dairy farms (Mezo et al., 2011). Currently, an "economic threshold" above which significant economic losses occur has been calculated to relate to a prevalence of $\geq$25% (Vercruysse and Claerebout, 2001). This percentage exceeds the 12% prevalence required to detect infected herds using bulk milk samples using a commercial test kit (MM3-SERO kit; cf. Section 5.4) (Mezo et al., 2010b), suggesting that the testing of such samples might assist in the early detection of infected herds and in the prediction of economic losses. However, another study (Ellis et al., 2011), using a commercial kit (Institut Pourquier, Montpellier, France), proposed that a prevalence of 50% of *F. hepatica* was necessary to detect infection in herds using bulk milk samples, whereas 20% was indicated by other authors using the same kit (Duscher et al., 2011). In addition, a seasonality in the presence of antibodies against *F. hepatica* in milk has been described and is important to consider when interpreting results (Kuerpick et al., 2012). Moreover, in ELISA, an optical density ratio of >0.8 has been estimated as an economic threshold for infection detection (Bennema et al., 2009). Although some authors have found a significant correlation in serological results between serum and milk samples (Mezo et al., 2010a,b; Salimi-Bejestani et al., 2007), low test specificity has been reported (Salimi-Bejestani et al., 2007), indicating that such tests can only be used for the screening of herds of cows for *F. hepatica* infection or fascioliasis. Based on this information, the value of bulk milk testing for the detection of infected herds and prediction of economic losses is controversial.

5.4. Commercially available diagnostic kits

Originally, an ELISA was developed for the detection of specific serum antigen called MM3 (Mezo et al., 2003). Subsequently, an indirect antibody-detection test was established using this antigen and developed commercially by Bio-X Diagnostics (La Jemelle, Belgium). Muino et al. (2011) showed that this antigen comprised cathepsins L1 and L2 and a Kunitz-like protein. According to Mezo et al. (2007), the test was able to detect serum antibodies in sheep infected with small numbers (5–40) of metacercariae. Seroconversion was observed 4 WPI in sheep infected with 1–2 flukes and 3 WPI in sheep infected with 3 flukes. Preliminary results also suggested that this test could be used for the detection of anti-*F. hepatica* antibodies in serum and milk from other ruminants (Mezo et al., 2007). The Bio-X ELISA kit has been shown to achieve similar performance for sheep sera compared with an ELISA using ES antigens (Salimi-Bejestani et al., 2005).

The fraction 2 (F2) antigen has received considerable attention as a diagnostic antigen (Biguet et al., 1962; Capron et al., 1965) and provided the basis for a commercial test kit for cattle (Pourquier, Montpellier, France). Initially, the F2 antigen was purified from *F. hepatica* (Tailliez and Korach, 1970) and used by Levieux et al. (1992a) in an immunohaemagglutination assay for the detection of *F. hepatica* infection in experimentally and naturally infected cattle. Using this initial test, antibodies were detected 2–4 WPI. Subsequently, these authors described an improved haemagglutination test (Levieux et al., 1992b) and used it for the detection of serum antibodies between the 2 and 3 WPI of *F. hepatica* infection in experimental goats (Levieux and Levieux, 1994). Reichel (2002) then used this test for the detection of serum antibodies in infected cattle from 2 WPI. However, no correlation was found between specific antibody levels and the intensity of *F. hepatica* infection. In contrast, Kuerpick et al. (2013) described that serum antibody levels detected in calves using this assay were significantly correlated to the infective dose.

Problems in the differentiation between positive and negative test results in sheep have confirmed the manufacturer's recommendation to use this kit exclusively for cattle. Rapsch et al. (2006) used the Pourquier kit to estimate the prevalence of fascioliasis in cattle at an abattoir. The sensitivity of this test was 91.7%, which is higher than that obtained by coprology (69%) and classical meat inspection for fascioliasis (63.2%).

6. DETECTION OF *F. hepatica* ANTIGENS

Because the detection of anti-*F. hepatica* serum or milk antibodies does not necessarily allow the differential diagnosis of current and past (cured) infections, the detection of specific antigens might be useful for the early detection of *F. hepatica* infection. In general, antigens circulating in serum are found from 1 WPI, much earlier than the detection of antigens in faeces around 4 WPI (Dumenigo et al., 1999, 2000; Espino et al., 1997a). Also, and in contrast to the levels of eggs detected by microscopy in faeces, a positive correlation has been reported between antigen levels and the parasite burden (Aygun and Baskaya, 1939; Espino et al., 1998; Mezo et al., 2004; Wagner, 1935). A summary of the time after infection when different antigens are detected in serum and faeces from different hosts experimentally infected with *F. hepatica* is given in Tables 2.4 and 2.5, respectively.

6.1. Detection of antigens circulating in the blood stream

Early attempts to detect circulating antigen F2 of *F. hepatica* were described by Robert et al. (1980) using a haemagglutination inhibition assay. In general, circulating antigens could be detected from 1 WPI in different animal species, and they could be detected for a period as long as 12 WPI in sheep using a sandwich ELISA based on polyclonal antibodies against ES products (Almazan et al., 2001). Langley and Hillyer (1989) detected serum antigens at 1 WPI and antibodies between 2 and 3 WPI in the same animals, suggesting that specific antigen detection could be used for early diagnosis. The antigens detected were shown to be *F. hepatica*-specific in experiments using sera from sheep infected with *Schistosoma mansoni* (see Rodriguez-Perez and Hillyer, 1995). Sanchez-Andrade et al. (2000) developed a sandwich ELISA using polyclonal rabbit anti-*F. hepatica* to detect a an ES antigen, with a specificity of 100% and a sensitivity of 86% in cattle from an endemic

Table 2.4 Summary of detection times of specific antigens in sera from different animals after experimental infection with *Fasciola hepatica*

Species	Antigen	Antigen detection (WPI)	Detection level (ng/ml)	References
Mice	ES	1–5	0.25	Langley and Hillyer (1989)
Sheep	ES	1–10	N/A	Arias et al. (2006)
Cattle	ES	1	N/A	Leclipteux et al. (1998)
Rat	ES ES78	1 3	45 880	Espino et al. (1997a)
Sheep	ES ES78	1–5	N/A	Dumenigo et al. (1999, 2000)
Sheep	ES ES78	1–12	190	Almazan et al. (2001)
Sheep	Cathepsin L1	3	N/A	Mezo et al. (2003)

ES, excretory/secretory; WPI, week(s) post infection; N/A, Not available.

Table 2.5 Summary of detection times of specific antigens in faeces from different animals experimentally infected with *Fasciola hepatica*

Species	Antigen	Antigen detection (WPI)	Detection level	References
Mice	ES	3–4	25 ng	Langley and Hillyer (1989)
Rat	ES78	6	95 ng/ml	Espino et al. (1997a)
Rat	Tegument	1–17	N/A	Paz-Silva et al. (2002)
Cattle	26–28 kDa	6	300 pg/ml	Abdel-Rahman et al. (1998)
Mice	Somatic	4	N/A	Moustafa et al. (1998)
Rats	Somatic	6	N/A	Moustafa et al. (1998)
Rabbit	Somatic	7	N/A	Moustafa et al. (1998)
Rat	ES78	4–16	95–730 ng/ml	Espino et al. (1997a)
Sheep	ES78	5–14WPI	N/A	Dumenigo et al. (1999, 2000)
Sheep	ES	4–12	90 ng/ml	Almazan et al. (2001)
Cattle	16–28 kDa	6	300 pg/ml	Abdel-Rahman et al. (1998)
Sheep	MM3	7–18	0.3 ng/ml	Mezo et al. (2004)
Sheep	MM3	6–34	N/A	Valero et al. (2009)

ES, excretory/secretory; WPI, week(s) post infection; N/A, Not available.

area; the results showed that less animals had circulating *F. hepatica* antigens (37.3%) than serum antibodies (85.1%). The same test was used under field conditions to assess prevalence in naturally infected sheep (Paz-Silva et al., 2003) and cattle (Arias et al., 2010). Espino et al. (1990), working in Cuba, developed a sandwich ELISA, aiming to detect circulating antigen using the monoclonal antibody (mAb) ES78, which has been described to bind a non-glycosylated antigen from fractions of the ES products (14, 24, 26 and 51 kDa) (Espino et al., 2000). This test, known as FasciDIG® ("Pedro Kouri" Institute of Tropical Medicine), has been used to test serum samples and can detect a minimum of 10 ng/ml of antigen (Espino et al., 1990). Recently, a minimum antigen detection threshold of 1.95 ng/ml has been reported for FasciDIG (Marcet Sanchez et al., 2012). By mixing known quantities of *F. hepatica* ES proteins to mouse serum, a detection limit of 25 and 0.25 ng/ml of these antigens was established using horse radish peroxidase and a biotin-labelled antibody, respectively (Langley and Hillyer, 1989). However, in practice, the utility of the biotinylated antibody was limited due to a high background in the assay, such that the actual detection limit in serum was estimated at 25 ng/ml. The antigen FhrAPS was detected in 32% of naturally infected cattle, suggesting ongoing infection (Arias et al., 2010). In contrast, 65% of those animals had antibodies against *F. hepatica*, which would include animals with both ongoing and past infections (Arias et al., 2010).

6.2. Specific detection of *F. hepatica* antigens in faeces (coproantigens)

A significant correlation has been found among the presence of coproantigens, egg output and intensity of infection (Abdel-Rahman et al., 1998; Almazan et al., 2001; Dumenigo et al., 1999, 2000). Such antigens also remain present in faeces for a longer period of time than *F. hepatica* antigens in serum (Pelayo et al., 1998). ES antigens from adult *F. hepatica*, characterized by Sanchez-Andrade et al. (2000), were detected in the faeces from experimentally infected rats from 1 WPI and decreasing at 17 WPI (Paz-Silva et al., 2002). The antigen recognized by mAb ES78 (FasciDIG®) has also been detected in the faeces from infected patients, with a detection limit of 15 ng/ml of faecal supernatant, and does not appear to cross-react with antigens from other parasites (Espino and Finlay, 1994). The test has also been applied to faeces from *F. hepatica*-infected cattle (Castro et al., 1994; Godoy et al., 2010), rats (Espino et al., 1997a), sheep (Almazan et al., 2001; Dumenigo and Finlay, 1998; Dumenigo et al., 1999, 2000) and alpacas (Li et al., 2005). The ES78-based coproantigen test, combined with the detection of an antigen in blood, allowed the specific detection of 91% of patients with clinical signs consistent with fascioliasis in a fascioliasis outbreak in Cuba (Espino et al., 1998). In a different study by the same research team, coproantigens were detected in all 19 patients chronically affected by fascioliasis. A correlation was found between the number of eggs and the amount of coproantigen and also between the number of eggs and level of immune complexes in blood (Pelayo et al., 1998).

Based on a chromatographic analysis of ES products (Mezo et al., 2003), Mezo et al. (2004) developed a mAb against the MM3 antigen (MM3/IgG1κ) that was used in a capture ELISA for the detection of this antigen in the faeces from sheep or cattle. It allowed the early detection of infection in sheep (3–5 WPI), including animals that received only 10 metacercariae of *F. hepatica* (Mezo et al., 2003). The antigen detection limits in this test were 0.3 and 0.6 ng/ml of *F. hepatica* ES antigens for sheep and cattle, respectively, which equates to detecting an infection in sheep and cattle with one and two adult *F. hepatica* parasites, respectively (Mezo et al., 2004). The first detection of *F. hepatica*-specific coproantigens in the MM3 capture

ELISA preceded the initial identification of eggs in faeces by 1–5 weeks (Mezo et al., 2003, 2004). However, in contrast to experimental studies, antigens were detected in the faeces from naturally infected lambs at the same time as eggs (Gordon et al., 2012). The test gave negative results using samples from both sheep and cattle without *F. hepatica* infection or infected with other parasites (including various species of nematodes and cestodes), confirming its specificity (Mezo et al., 2004). The commercial kit based on the MM3 antigen (Bio K201 ELISA, **Bio-X Diagnostics) was also shown to be highly specific (100%), with negative results when tested on soluble fractions of homogenates from *Paramphistomum cervi* and *T. hydatigena* (Kajugu et al., 2012). Ubeira et al. (2009) detected the MM3 antigen in 23 stool samples containing *F. hepatica* eggs, while no antigens were detected in 213 samples from patients with no detectable parasitic infections. The MM3-COPRO test was also evaluated in the field in two endemic regions of Bolivia and Peru, achieving similar results to the microscopic examination of faeces for eggs (which has limited sensitivity for the detection of animals excreting low numbers of *Fasciola* eggs). Therefore, given the similarity in results, the MM3-COPRO test will likely not detect animals excreting low numbers of eggs in faeces (Valero et al., 2012a).

A previously described antigen of *F. hepatica* (26–28 kDa) from the tegument and gut cells of this parasite (Abdel-Rahman et al., 1999) was also used as a target for a capture ELISA, identifying 13 calves infected with $\geq$10 adults in the liver. In this case, the minimum detection level of the 26–28 kDa antigen was 300 pg/ml of faecal supernatants (Abdel-Rahman et al., 1998; el-Bahi et al., 1992).

6.3. Specific detection of antigens in bile

Somatic antigens (27.5 kDa) have been detected using polyclonal serum-based immunodiffusion in the bile from rats, rabbits or sheep heavily infected with *F. hepatica* (see Klimenko, 1980), but were not detected in serum from these animals. Other antigens include ES products of *F. hepatica* for the detection of a 26 kDa antigen in the bile of *F. hepatica*-infected cattle by immunoblotting (el-Bahi et al., 1992).

6.4. Specific detection of antigens following anthelmintic treatment

In general, serum antibodies in animals tend to persist for prolonged periods of time following the resolution of an infection, whereas antigen levels recede rapidly following anthelmintic treatment (Hillyer, 1998). Hence, there has been considerable effort to determine when antigens are no longer detectable after treatment. In a study performed in an endemic area in Peru, Knobloch (1985) defined a patient as cured when no *F. hepatica* antigen was detectable 12 WPT. Espino et al. (1992) found no circulating antigens in people 12 WPT with bithionol. Similar results were obtained in another study, in which no ES78 antigens were detected in people 8 WPT (Espino and Finlay, 1994). Hammouda et al. (1997) showed that the level of specific antigens decreased in all patients 12 WPT. Subsequently, Sanchez-Andrade et al. (2001) demonstrated reduced levels of circulating antigen in sheep 2–4 WPT with triclabendazole using a previously described test (Sanchez-Andrade et al., 2000). More recently, using the Bio K201 ELISA, no MM3 antigen was detected in faeces from sheep 7 days post treatment with triclabendazole (Novobilsky et al., 2012), much earlier than the 2 WPT determined in previous experiments (Flanagan et al., 2011a,b).

Effective treatment of fascioliasis has been defined as the absence of *Fasciola* antigens in faeces 2 WPT using a coproantigen reduction test (CRT) (Flanagan et al., 2011a). Flanagan et al. (2011b) standardized a CRT for the diagnosis of resistance to triclabendazole in *F. hepatica* based on the use of the Bio K201 kit. Sheep infected with a susceptible strain of *F. hepatica* (Cullompton) excreted no parasite antigens 2 WPT. On the other hand, such antigens persisted 4 WPT in sheep infected with a resistant strain of *F. hepatica* (Sligo). The same kit was also used by Gordon et al. (2012) to independently assess its performance under field conditions; there was a significant decrease in coproantigens in sheep from 1 WPT.

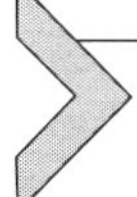

7. DNA METHODS FOR THE GENETIC IDENTIFICATION AND CHARACTERIZATION OF *Fasciola* SPECIES AND THE SPECIFIC DIAGNOSIS OF FASCIOLIASIS

7.1. Molecular methods for the genetic characterization of species of *Fasciola*

To date, various PCR-based methods, utilizing nuclear and/or mitochondrial DNA targets, have been applied for the genetic characterization, identification and/or differentiation of *Fasciola* spp. (e.g. Ai et al., 2010b, 2011; Alasaad et al., 2011a,b; Caron et al., 2011; Ichikawa and Itagaki, 2010; Le et al., 2012a,b; McGarry et al., 2007). For example, fingerprinting approaches, such as random amplification of polymorphic DNA (rAPS) (Aldemir, 2006) and microsatellite analysis (Hurtrez-BoussÈS et al., 2004), have been employed to assess polymorphism within and between *F. hepatica* populations (Semyenova et al., 2003; Vargas et al., 2003). Also, analyses of mitochondrial gene regions and internal transcribed spacers (ITSs) have characterized multiple genetic variants of *F. hepatica*, *F. gigantica* and intermediate forms (predicted to be hybrids) (Mas-Coma et al., 2009). Moreover, recently, mitochondrial DNA sequence analyses were conducted within and between TCBZ-resistant and TCBZ-susceptible populations of *F. hepatica* in Australia, with the findings indicating limited genetic variation between resistant and nonresistant populations (Elliott et al., 2013).

Using fluorescence-based PCR-linked mutation scanning analysis, Alasaad et al. (2011b) were able to characterize and differentiate *F. hepatica*, *F. gigantica* and an intermediate form of *Fasciola* originating from different definitive hosts (horse, sheep and cattle) in China, Spain, Nigeria and Egypt. In another study, this team (Alasaad et al., 2011a) also developed a quantitative real-time PCR (qPCR) using markers in ITS-2 and species-specific internal *Taq*Man probes, to differentiate *Fasciola* species. More recently, Ai et al. (2010b) established a loop-mediated isothermal amplification (LAMP) assay for the specific identification of *F. hepatica* and *F. gigantica* using four species-specific primer sets. The authors concluded that the LAMP assay was more sensitive than the conventional, specific PCR assays, consistent with some other studies of other parasites (Karanis et al., 2007; Kuboki et al., 2003; Le et al., 2012b).

7.2. Specific detection of *F. hepatica* DNA in faeces

There are only a few reports of the use molecular tools for the detection of *F. hepatica* DNA in faeces. For example, Kozak et al. (2008) described the use of PCR to amplify a 124 bp tandem repeat sequence from faecal samples from experimentally infected rats, sheep and cattle. This PCR assay detected infected rats 5 WPI, with 100% of the animals being test-positive at 9 WPI. In sheep, PCR test-positive results were obtained at 8 WPI, while in cattle, *F. hepatica* DNA was detected from 10 WPI. In all cases, DNA was specifically detected in faeces by PCR before eggs were detected by microscopic examination. In another study, Le et al. (2012a) used a duplex PCR for the detection of *F. hepatica* and *F. gigantica* mitochondrial DNA in snails and in faecal samples from cattle, goats, buffaloes and sheep. Here, again, PCR was able to detect infection well before the microscopic detection of eggs, which eventually confirmed *F. hepatica* infection.

Recently, Martinez-Perez et al. (2012) showed that a nested PCR was considerably more sensitive than conventional faecal examination and a commercial immunoassay for the diagnosis of *F. hepatica* infection in experimentally infected sheep. Specifically, through the amplification of a 423 bp mitochondrial DNA region, the PCR assay was able to detect infected animals 3 WPI, whereas a nested PCR-based assay could bring this forwards to 2 WPI. In Pakistan, a PCR-based assay amplifying the ITS-2 region was used to assess the prevalence of *F. hepatica* infection in 200 faecal and bile samples from sheep and goats, and compared findings to those achieved by microscopic detection of eggs. In sheep, *F. hepatica* DNA was detected in 4.5% and 8.5% of faecal and bile samples, whereas eggs were detected microscopically in only 3.5% and 0.5% of these samples, respectively. In goats, such DNA was detected in 3.5% and 5% of all faecal and bile samples, whereas eggs were detected in 2% and 4% of the same samples, respectively (Waseem et al., 2012).

7.3. Molecular detection of *F. hepatica* in snails

Although the main focus of this chapter is on diagnosis in the definitive mammalian (definitive) host, the genetic identification, characterization and differentiation of larval stages of *F. hepatica* and related species in snail intermediate hosts are central to studying the epidemiology of fascioliasis. Early attempts at detecting infection in snails were conducted using hybridization employing repetitive DNA probes (e.g. Heussler et al., 1993). Subsequently, numerous distinct PCR-coupled approaches have been used for detection and characterization, employing genetic markers principally in nuclear ribosomal DNA (including rRNA genes and intervening ITSs) or mitochondrial genes, which usually achieved detection limits of ≥ 1 pg parasite DNA (e.g. Ai et al., 2010a; Caron et al., 2011; Cucher et al., 2006; Le et al., 2012b; Magalhaes et al., 2004, 2008; Marcilla et al., 2002; Martinez-Ibeas et al., 2013; McGarry et al., 2007; Rognlie et al., 1994; Rokni et al., 2010).

8. CONCLUDING REMARKS

Despite relatively intensive research, there still is no highly specific and sensitive method for the diagnosis of *F. hepatica* or *F. gigantica* infection/fascioliasis and the differentiation of pre-patent from patent infection. This chapter covers the advantages and disadvantages of methods ranging from the microscopic detection of *F. hepatica* eggs in faeces, to immunological diagnostic methods, to more modern techniques, such as molecular tools for the specific detection of *Fasciola* DNA (Fig. 2.1). A summary of advantages and disadvantages of the diagnostic methods is found in Table 2.6. The

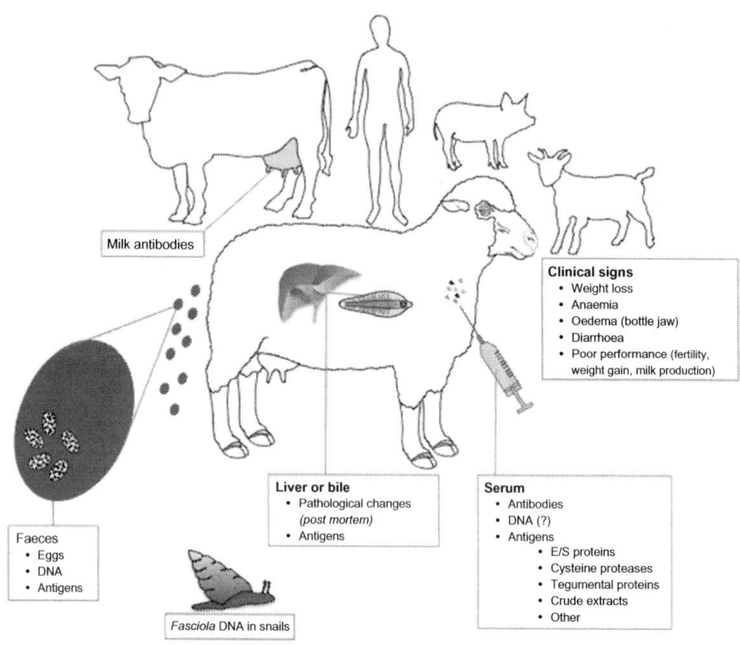

Test	Advantages	Disadvantages	Detection after infection	Distinguishes current from past infection
Microscopic egg detection	Inexpensive	Low sensitivity	From 8 WPI	Detects patent infection
Antibody detection in sera		Cross-reactivity No correlation with parasite burden Cost of ELISA reader	From 2 WPI	No
Antibody detection in milk	Can estimate losses due to infection/exposure to the parasite	Cross-reactivity Cost of ELISA reader		No
Antigen detection in sera	Positive correlation with parasite burden	Only detects migrating stages in liver Cost of ELISA reader	From 2 WPI	Detects current infection
Antigen detection in faeces	Positive correlation with parasite burden	Cost of ELISA reader	From 4 WPI	Detects patent infection
DNA detection (PCR)	Highly sensitive and specific	Cost of thermocycler		Detects patent infection

Figure 2.1 Schematic representation of the main diagnostic methods used for the diagnosis of *Fasciola hepatica* infection/fascioliasis. Some of the advantages and disadvantages of these methods are listed in the accompanying table. WPI, weeks post infection.

Table 2.6 Advantages and disadvantages of the currently available diagnostic tests for *Fasciola hepatica*

Test	Advantages	Disadvantages	Detection after infection	Discriminate current infection
Microscopic egg detection	Inexpensive	Low sensitivity	From 8 WPI	Yes
Antibody detection in sera		Cross-reactivity No correlation with parasite burden	From 2 WPI	No
Antibody detection in milk	Easy to perform Can estimate losses due to infection/exposure to the parasite	Cross-reactivity		No
Antigen detection in sera	Positive correlation with parasite burden	Only detects migrating stages in liver	From 2 WPI	Yes
Antigen detection in faeces	Positive correlation with parasite burden		From 4 WPI	Yes
PCR	Highly sensitive and specific			Yes, only when adults parasites are present

WPI, weeks post infection.

molecular tools seem to have considerable promise for the diagnosis using stool samples from animals (Fayer et al., 2012; Mejia et al., 2013; Roeber et al., 2012; Stark et al., 2011; Taniuchi et al., 2011). Nonetheless, specific diagnosis in animals infected with juvenile or immature worms that do not shed eggs into the biliary system might remain a challenge. It might be possible to detect circulating *F. hepatica* DNA or RNA in blood, as has been suggested for the diagnosis of *E. granulosus* (see McManus, 2014), but such an approach remains to be critically evaluated. As there is no universally applicable technique for the effective isolation of tiny amounts of nucleic acids of *Fasciola* from the different biological matrices (e.g. blood, faeces or bile) or for the removal of substances inhibitory to the PCR-based amplification process, techniques need to be rigorously assessed. Therefore, despite some recent advances, there is a need for practical and cost-effective assays for the specific diagnosis of both pre-patent and patent *F. hepatica* and *F. gigantica* infections, to complement or replace traditional methods used at this point in time.

Major progress in next-generation nucleic acid sequencing technologies (Koboldt et al., 2013) also provides exciting new prospects for improved diagnostic and analytic applications. Following the rapid reduction in cost of sequencing, these technologies have become accessible to many research groups and now enable the direct sequencing of parasite genomic DNA derived from biological matrices (such as faeces, blood and milk), followed by bioinformatic identification and extraction of specific sequences for diagnosis and genetic characterization. These advanced technologies might change the face of molecular diagnostics in the future.

Clearly, accurate diagnosis is central to (i) estimating the distribution and prevalence of *Fasciola* spp. and drug resistance (combined with the classical method of detecting resistance by faecal egg count reduction; Daniel et al., 2012; Martinez-Valladares et al., 2014); (ii) assessing alterations in prevalence and geographic distribution following anthelmintic treatment; and (iii) establishing effective treatment and control of fascioliasis. Given the emergence of drug resistance in *F. hepatica* populations in some countries (Alvarez-Sanchez et al., 2006; Moll et al., 2000; Ortiz et al., 2013; Overend and Bowen, 1995), the characterization of molecular markers associated with drug resistance (Hodgkinson et al., 2013) would be important, so that they also could be incorporated into a species- or genus-specific diagnostic assay. The establishment of highly effective molecular-diagnostic tools will have major implications for the management of fascioliasis in animals.

ACKNOWLEDGEMENTS

Funding from the Australian Research Council (ARC) is gratefully acknowledged. Other support from the Australian Academy of Science, the Australian-American Fulbright Commission, Alexander von Humboldt Foundation, and Melbourne Water Corporation (R. B. G.) is gratefully acknowledged, as is support from the Victorian Life Sciences Computation Initiative (VLSCI) grant number VR0007 on its Peak Computing Facility at the University of Melbourne, an initiative of the Victorian Government (R. B. G.).

REFERENCES

Abdel-Rahman, S.M., O'Reilly, K.L., Malone, J.B., 1998. Evaluation of a diagnostic monoclonal antibody-based capture enzyme-linked immunosorbent assay for detection of a 26- to 28-kd *Fasciola hepatica* coproantigen in cattle. Am. J. Vet. Res. 59, 533–537.

Abdel-Rahman, S., O'Reilly, K.L., Malone, J.B., 1999. Biochemical characterization and localization of *Fasciola hepatica* 26–28 kDa diagnostic coproantigen. Parasit. Immunol. 21, 279–286.

Adela Valero, M., Perez-Crespo, I., Victoria Periago, M., Khoubbane, M., Mas-Coma, S., 2009. Fluke egg characteristics for the diagnosis of human and animal fascioliasis by *Fasciola hepatica* and *F. gigantica*. Acta Trop. 111, 150–159.

Ai, L., Dong, S.J., Zhang, W.Y., Elsheikha, H.M., Mahmmod, Y.S., Lin, R.Q., Yuan, Z.G., Shi, Y.L., Huang, W.Y., Zhu, X.Q., 2010a. Specific PCR-based assays for the identification of *Fasciola* species: their development, evaluation and potential usefulness in prevalence surveys. Ann. Trop. Med. Parasitol. 104, 65–72.

Ai, L., Li, C., Elsheikha, H.M., Hong, S.J., Chen, J.X., Chen, S.H., Li, X., Cai, X.Q., Chen, M.X., Zhu, X.Q., 2010b. Rapid identification and differentiation of *Fasciola hepatica* and *Fasciola gigantica* by a loop-mediated isothermal amplification (LAMP) assay. Vet. Parasitol. 174, 228–233.

Ai, L., Chen, M., Alasaad, S., Elsheikha, H.M., Li, J., Li, H., Lin, R., Zou, F., Zhu, X., Chen, J., 2011. Genetic characterization, species differentiation and detection of *Fasciola* spp. by molecular approaches. Parasit. Vectors 4, 101.

Alasaad, S., Soriguer, R.C., Abu-Madi, M., El-Behairy, A., Jowers, M.J., Banos, P.D., Piriz, A., Fickel, J., Zhu, X., 2011a. A TaqMan real-time PCR-based assay for the identification of *Fasciola* spp. Vet. Parasitol. 179, 266–271.

Alasaad, S., Soriguer, R.C., Abu-Madi, M., El Behairy, A., Banos, P.D., Piriz, A., Fickel, J., Zhu, X.Q., 2011b. A fluorescence-based polymerase chain reaction-linked single-strand conformation polymorphism (F-PCR-SSCP) assay for the identification of *Fasciola* spp. Parasitol. Res. 108, 1513–1517.

Aldemir, O.S., 2006. Differentiation of cattle and sheep originated *Fasciola hepatica* by RAPD-PCR. Rev. Med. Vet.-Toulouse 157, 65–67.

Almazan, C., Avila, G., Quiroz, H., Ibarra, F., Ochoa, P., 2001. Effect of parasite burden on the detection of *Fasciola hepatica* antigens in sera and feces of experimentally infected sheep. Vet. Parasitol. 97, 101–112.

Alvarez-Sanchez, M.A., Mainar-Jaime, R.C., Perez-Garcia, J., Rojo-Vazquez, F.A., 2006. Resistance of *Fasciola hepatica* to triclabendazole and albendazole in sheep in Spain. Vet. Rec. 159, 424–425.

Ambrorse-Thomas, P., Desgeorges, P.T., Bouttaz, M., 1980. ELISA test for detecting circulating antibody and/or antigen in human and bovine fascioliasis. Ann. Soc. Belg. Med. Trop. 60, 47–60.

Anderson, P.H., Berrett, S., Brush, P.J., Hebert, C.N., Parfitt, J.W., Patterson, D.S., 1977. Biochemical indicators of liver injury in calves with experimental fascioliasis. Vet. Rec. 100, 43–45.

Andrews, S.J., 1999. The life cycle of *Fasciola hepatica*. In: Dalton, J.P. (Ed.), Fasciolosis. CABI, Oxon, UK, pp. 1–30.

Arias, M., Hillyer, G.V., Sanchez-Andrade, R., Suarez, J.L., Pedreira, J., Lomba, C., Diaz, P., Morrondo, P., Diez-Banos, P., Paz-Silva, A., 2006. A 2.9 kDa *Fasciola hepatica*-recombinant protein based ELISA test for the detection of current-ovine fasciolosis trickle infected. Vet. Parasitol. 137, 67–73.

Arias, M., Morrondo, P., Hillyer, G.V., Sanchez-Andrade, R., Suarez, J.L., Lomba, C., Pedreira, J., Diaz, P., Diez-Banos, P., Paz-Silva, A., 2007. Immunodiagnosis of current fasciolosis in sheep naturally exposed to *Fasciola hepatica* by using a 2.9 kDa recombinant protein. Vet. Parasitol. 146, 46–49.

Arias, M.S., Suarez, J.L., Hillyer, G.V., Francisco, I., Calvo, E., Sanchez-Andrade, R., Diaz, P., Francisco, R., Diez-Banos, P., Morrondo, P., Paz-Silva, A., 2009. A recombinant-based ELISA evaluating the efficacy of netobimin and albendazole in ruminants with naturally acquired fascioliasis. Vet. J. 182, 73–78.

Arias, M., Pineiro, P., Hillyer, G.V., Suarez, J.L., Francisco, I., Cortinas, F.J., Diez-Banos, P., Morrondo, P., Sanchez-Andrade, R., Paz-Silva, A., 2010. An approach of the laboratory to the field: assessment of the influence of cattle management on the seroprevalence of fascioliasis by using polyclonal- and recombinant-based ELISAs. J. Parasitol. 96, 626–631.

Arias, M.S., Pineiro, P., Hillyer, G.V., Francisco, I., Cazapal-Monteiro, C.F., Suarez, J.L., Morrondo, P., Sanchez-Andrade, R., Paz-Silva, A., 2012. Enzyme-linked immunosorbent assays for the detection of equine antibodies specific to a recombinant *Fasciola hepatica* surface antigen in an endemic area. Parasitol. Res. 110, 1001–1007.

Aygun, S.T., Baskaya, H., 1939. Anwendung der Allergie-Reaktion bei der Bekampfung der Distomatose. Tierarztliche Rundschau 45, 379–382.

Bautista-Garfias, C.R., Lopez-Arellano, M.E., Sanchez-Albarran, A., 1989. A new method for serodiagnosis of sheep fascioliasis using helminth excretory–secretory products. Parasitol. Res. 76, 135–137.

Behm, C., Sangster, N., 1999. Pathology, pathophysiology and clinical aspects. In: Dalton, J.P. (Ed.), Fasciolosis. UK CABI Publishing, Oxon, pp. 185–224.

Benex, J., Lamy, L., Gledel, J., 1959. Complement-fixation test for liver flukes in sheep. Bull. Soc. Pathol. Exot. 52, 83–87.

Bennema, S., Vercruysse, J., Claerebout, E., Schnieder, T., Strube, C., Ducheyne, E., Hendrickx, G., Charlier, J., 2009. The use of bulk-tank milk ELISAs to assess the spatial distribution of *Fasciola hepatica, Ostertagia ostertagi* and *Dictyocaulus viviparus* in dairy cattle in Flanders (Belgium). Vet. Parasitol. 165, 51–57.

Bennett, C.E., Joshua, G.W., Hughes, D.L., 1982. Demonstration of juvenile-specific antigens of *Fasciola hepatica*. J. Parasitol. 68, 791–795.

Berasain, P., Goni, F., McGonigle, S., Dowd, A., Dalton, J.P., Frangione, B., Carmona, C., 1997. Proteinases secreted by *Fasciola hepatica* degrade extracellular matrix and basement membrane components. J. Parasitol. 83, 1–5.

Berasain, P., Carmona, C., Frangione, B., Dalton, J.P., Goni, F., 2000. *Fasciola hepatica*: parasite-secreted proteinases degrade all human IgG subclasses: determination of the specific cleavage sites and identification of the immunoglobulin fragments produced. Exp. Parasitol. 94, 99–110.

Biguet, J., Capron, A., Tran Van Ky, P., 1962. Antigens of *Fasciola hepatica*: electro-phoretic and immuno-electrophoretic study, and a comparison with antigens from seven other helminths. Ann. Parasit. Humaine Comparee. 37, 221–231.

Boray, J.C., 1969. Experimental fascioliasis in Australia. Adv. Parasitol. 7, 95–210.

Boray, J.C., Crowfoot, P.D., Strong, M.B., Allison, J.R., Schellenbaum, M., Von Orelli, M., Sarasin, G., 1983. Treatment of immature and mature *Fasciola hepatica* infections in sheep with triclabendazole. Vet. Rec. 113, 315–317.

Bossaert, K., Farnir, F., Leclipteux, T., Protz, M., Lonneux, J.F., Losson, B., 2000. Humoral immune response in calves to single-dose, trickle and challenge infections with *Fasciola hepatica*. Vet. Parasitol. 87, 103–123.

Boulard, C., Carreras, F., Van Gool, F., 1995. Evaluation of nitroxynil and closantel activity using ELISA and egg counts against *Fasciola hepatica* in experimentally and naturally infected cattle. Vet. Res. 26, 249–255.

Brockwell, Y.M., Elliott, T.P., Anderson, G.R., Stanton, R., Spithill, T.W., Sangster, N.C., 2014. Confirmation of Fasciola hepatica resistant to triclabendazole in naturally infected Australian beef and dairy cattle. Int. J. Parasitol. Drugs Drug Resist. 4, 48–54.

Caban-Hernandez, K., Gaudier, J.F., Espino, A.M., 2012. Characterization and differential expression of a ferritin protein from *Fasciola hepatica*. Mol. Biochem. Parasitol. 182, 54–61.

Capron, A., Rose, G., Luffau, G., Biguet, J., Rose, F., 1965. Contribution of experimental distomiasis to the knowledge of human distomiasis caused by *Fasciola hepatica*. Immunological aspects. Rev. Immunol. Ther. Antimicrob. 29, 25–41.

Carmona, C., Dowd, A.J., Smith, A.M., Dalton, J.P., 1993. Cathepsin L proteinase secreted by *Fasciola hepatica* in vitro prevents antibody-mediated eosinophil attachment to newly excysted juveniles. Mol. Biochem. Parasitol. 62, 9–17.

Carnevale, S., Rodriguez, M.I., Santillan, G., Labbe, J.H., Cabrera, M.G., Bellegarde, E.J., Velasquez, J.N., Trgovcic, J.E., Guarnera, E.A., 2001. Immunodiagnosis of human fascioliasis by an enzyme-linked immunosorbent assay (ELISA) and a micro-ELISA. Clin. Diagn. Lab. Immunol. 8, 174–177.

Caron, Y., Righi, S., Lempereur, L., Saegerman, C., Losson, B., 2011. An optimized DNA extraction and multiplex PCR for the detection of *Fasciola* sp. in lymnaeid snails. Vet. Parasitol. 178, 93–99.

Carpenter, H.A., 1998. Bacterial and parasitic cholangitis. Mayo Clin. Proc. 73, 473–478.

Castro, J., Dumenigo, B., Espino, A., 1994. Detection of coproantigens for diagnosing active *Fasciola hepatica* infections in cattle. Parasitologia al Dia 18, 33–38.

Castro, E., Freyre, A., Hernandez, Z., 2000. Serological responses of cattle after treatment and during natural re-infection with *Fasciola hepatica*, as measured with a dot-ELISA system. Vet. Parasitol. 90, 201–208.

Cervi, L.A., Rubinstein, H., Masih, D.T., 1992. Serological, electrophoretic and biological properties of *Fasciola hepatica* antigens. Rev. Inst. Med. Trop. Sao Paulo 34, 517–525.

Chauvin, A., Moreau, E., Boulard, C., 1997. Diagnosis of bovine fascioliasis using serology of pools of sera. Interpretation in field conditions. Vet. Res. 28, 37–43.

Collins, P.R., Stack, C.M., O'Neill, S.M., Doyle, S., Ryan, T., Brennan, G.P., Mousley, A., Stewart, M., Maule, A.G., Dalton, J.P., Donnelly, S., 2004. Cathepsin L1, the major protease involved in liver fluke (*Fasciola hepatica*) virulence: propetide cleavage sites and autoactivation of the zymogen secreted from gastrodermal cells. J. Biol. Chem. 279, 17038–17046.

Cordova, M., Herrera, P., Nopo, L., Bellatin, J., Naquira, C., Guerra, H., Espinoza, J.R., 1997. *Fasciola hepatica* cysteine proteinases: immunodominant antigens in human fascioliasis. Am. J. Trop. Med. Hyg. 57, 660–666.

Cordova, M., Reategui, L., Espinoza, J.R., 1999. Immunodiagnosis of human fascioliasis with *Fasciola hepatica* cysteine proteinases. Trans. R. Soc. Trop. Med. Hyg. 93, 54–57.

Cornelissen, J.B., de Leeuw, W.A., van der Heijden, P.J., 1992. Comparison of an indirect haemagglutination assay and an ELISA for diagnosing *Fasciola hepatica* in experimentally and naturally infected sheep. Vet. Q. 14, 152–156.

Cornelissen, J.B., Gaasenbeek, C.P., Boersma, W., Borgsteede, F.H., van Milligen, F.J., 1999. Use of a pre-selected epitope of cathepsin-L1 in a highly specific peptide-based immunoassay for the diagnosis of *Fasciola hepatica* infections in cattle. Int. J. Parasitol. 29, 685–696.

Cornelissen, J.B., Gaasenbeek, C.P., Borgsteede, F.H., Holland, W.G., Harmsen, M.M., Boersma, W.J., 2001. Early immunodiagnosis of fasciolosis in ruminants using recombinant *Fasciola hepatica* cathepsin L-like protease. Int. J. Parasitol. 31, 728–737.

Corvo, I., Cancela, M., Cappetta, M., Pi-Denis, N., Tort, J.F., Roche, L., 2009. The major cathepsin L secreted by the invasive juvenile *Fasciola hepatica* prefers proline in the S2 subsite and can cleave collagen. Mol. Biochem. Parasitol. 167, 41–47.

Coudert, J., Garin, J.P., Ambroise-Thomas, P., Kien Troung, T., Despeignes, J., Pothier, M.A., 1967. An immune-fluorescence test on histological sections of *Fasciola hepatica*: a new technique for the serological diagnosis of fascioliasis. (First results). Bull. Soc. Pathol. Exot. 60, 71–79, +72 plates.

Cringoli, G., Rinaldi, L., Maurelli, M.P., Utzinger, J., 2010. FLOTAC: new multivalent techniques for qualitative and quantitative copromicroscopic diagnosis of parasites in animals and humans. Nat. Protoc. 5, 503–515.

Cucher, M.A., Carnevale, S., Prepelitchi, L., Labbe, J.H., Wisnivesky-Colli, C., 2006. PCR diagnosis of *Fasciola hepatica* in field-collected *Lymnaea columella* and *Lymnaea viatrix* snails. Vet. Parasitol. 137, 74–82.

Curasson, G., 1935. Recherches sur le diagnostic des distomatoses a *Fasciola hepatica* et *Amphistomum cervi* par les reactions allergiques. Bull. Acad. Vet. Fr. 8, 77–81.

Daniel, R., van Dijk, J., Jenkins, T., Akca, A., Mearns, R., Williams, D.J., 2012. Composite faecal egg count reduction test to detect resistance to triclabendazole in Fasciola hepatica. Vet. Rec. 171 (153), 151–155.

De Almeida, M.A., Ferreira, M.B., Planchart, S., Terashima, A., Maco, V., Marcos, L., Gotuzzo, E., Sanchez, E., Naquira, C., Scorza, J.V., Incani, R.N., 2007. Preliminary antigenic characterisation of an adult worm vomit preparation of *Fasciola hepatica* by infected human sera. Rev. Inst. Med. Trop. Sao Paulo 49, 31–35.

de Weil, N.S., Hillyer, G.V., Pacheco, E., 1984. Isolation of *Fasciola hepatica* genus-specific antigens. Int. J. Parasitol. 14, 197–206.

Deelder, A.M., Ploem, J.S., 1975. An immunofluorescence reaction for *Fasciola hepatica* using the defined antigen substrate spheres (DASS) system. Exp. Parasitol. 37, 173–178.

Dorsman, W., 1960. The diagnosis of sub-clinical fascioliasis by means of faecal examination, and the control of liver flukes (*Fasciola hepatica*). Bull. Off. Int. Epizoot. 54, 502–508.

Dumenigo, B.E., Finlay, C.M., 1998. Detection and quantification of *Fasciola hepatica* antigens in sheep. Rev. Cubana Med. Trop. 50, 82–84.

Dumenigo, B.E., Espino, A.M., Finlay, C.M., Mezo, M., 1999. Kinetics of antibody-based antigen detection in serum and faeces of sheep experimentally infected with *Fasciola hepatica*. Vet. Parasitol. 86, 23–31.

Dumenigo, B.E., Espino, A.M., Finlay, C.M., Mezo, M., 2000. Kinetics of antibody-based antigen detection in serum and faeces of sheep experimentally infected with *Fasciola hepatica*. Vet. Parasitol. 89, 153–161.

Duscher, R., Duscher, G., Hofer, J., Tichy, A., Prosl, H., Joachim, A., 2011. *Fasciola hepatica*—monitoring the milky way? The use of tank milk for liver fluke monitoring in dairy herds as base for treatment strategies. Vet. Parasitol. 178, 273–278.

el-Bahi, M.M., Malone, J.B., Todd, W.J., Schnorr, K.L., 1992. Detection of stable diagnostic antigen from bile and feces of *Fasciola hepatica* infected cattle. Vet. Parasitol. 45, 157–167.

Elliott, T., Muller, A., Brockwell, Y., Murphy, N., Grillo, V., Toet, H.M., Anderson, G., Sangster, N., Spithill, T.W., 2013. Evidence for high genetic diversity of NAD1 and COX1 mitochondrial haplotypes among triclabendazole resistant and susceptible populations and field isolates of Fasciola hepatica (liver fluke) in Australia. Vet. Parasitol. 200, 90–96.

Ellis, K.A., Jackson, A., Bexiga, R., Matthews, J., McGoldrick, J., Gilleard, J., Forbes, A.B., 2011. Use of diagnostic markers to monitor fasciolosis and gastrointestinal nematodes on an organic dairy farm. Vet. Rec. 169, 524.

Espino, A.M., Finlay, C.M., 1994. Sandwich enzyme-linked immunosorbent assay for detection of excretory secretory antigens in humans with fascioliasis. J. Clin. Microbiol. 32, 190–193.

Espino, A.M., Hillyer, G.V., 2003. Molecular cloning of a member of the *Fasciola hepatica* saposin-like protein family. J. Parasitol. 89, 545–552.

Espino, A.M., Pico, M.C., Lopez, S., Dumenigo, B.E., Barbaru, D., Huesca, N., 1987. Partial purification and characterization of a *Fasciola hepatica* somatic antigen. Rev. Cubana Med. Trop. 39, 23–31.

Espino, A.M., Marcet, R., Finlay, C.M., 1990. Detection of circulating excretory secretory antigens in human fascioliasis by sandwich enzyme-linked immunosorbent assay. J. Clin. Microbiol. 28, 2637–2640.

Espino, A.M., Millan, J.C., Finlay, C.M., 1992. Detection of antibodies and circulating excretory-secretory antigens for assessing cure in patients with fascioliasis. Trans. R. Soc. Trop. Med. Hyg. 86, 649.

Espino, A.M., Marcet, R., Finlay, C.M., 1997a. *Fasciola hepatica*: detection of antigenemia and coproantigens in experimentally infected rats. Exp. Parasitol. 85, 117–120.

Espino, A.M., Padron, L., Dumenigo, B., Laferte, J., 1997b. Indirect ultra-micro-ELISA for detecting IgG antibodies in patients with fascioliasis. Rev. Cubana Med. Trop. 49, 167–173.

Espino, A.M., Diaz, A., Perez, A., Finlay, C.M., 1998. Dynamics of antigenemia and coproantigens during a human *Fasciola hepatica* outbreak. J. Clin. Microbiol. 36, 2723–2726.

Espino, A.M., Borges, A., Dumenigo, B.E., 2000. Fecal antigens of *Fasciola hepatica* potentially useful in the diagnosis of fascioliasis. Rev. Panam. Salud Publica 7, 225–231.

Espinoza, J.R., Timoteo, O., Herrera-Velit, P., 2005. Fas2-ELISA in the detection of human infection by *Fasciola hepatica*. J. Helminthol. 79, 235–240.

Espinoza, J.R., Maco, V., Marcos, L., Saez, S., Neyra, V., Terashima, A., Samalvides, F., Gotuzzo, E., Chavarry, E., Huaman, M.C., Bargues, M.D., Valero, M.A., Mas-Coma, S., 2007. Evaluation of Fas2-ELISA for the serological detection of *Fasciola hepatica* infection in humans. Am. J. Trop. Med. Hyg. 76, 977–982.

FAO, 1994. Diseases of Domestic Animals Caused By Flukes: Epidemiology, Diagnosis and Control of *Fasciola*, *Paramphistome*, *Dicrocoelium*, *Eurytrema* and *Schistosome* Infections of Ruminants in Developing Countries. FAO, Rome.

Farrell, C.J., Shen, D.T., Wescott, R.B., Lang, B.Z., 1981. An enzyme-linked immunosorbent assay for diagnosis of *Fasciola hepatica* infection in cattle. Am. J. Vet. Res. 42, 237–240.

Fayer, R., Santin, M., Macarisin, D., 2012. Detection of concurrent infection of dairy cattle with *Blastocystis*, *Cryptosporidium*, *Giardia*, and *Enterocytozoon* by molecular and microscopic methods. Parasitol. Res. 111, 1349–1355.

Flanagan, A., Edgar, H.W., Gordon, A., Hanna, R.E., Brennan, G.P., Fairweather, I., 2011a. Comparison of two assays, a faecal egg count reduction test (FECRT) and a coproantigen reduction test (CRT), for the diagnosis of resistance to triclabendazole in *Fasciola hepatica* in sheep. Vet. Parasitol. 176, 170–176.

Flanagan, A.M., Edgar, H.W., Forster, F., Gordon, A., Hanna, R.E., McCoy, M., Brennan, G.P., Fairweather, I., 2011b. Standardisation of a coproantigen reduction test (CRT) protocol for the diagnosis of resistance to triclabendazole in *Fasciola hepatica*. Vet. Parasitol. 176, 34–42.

Foreyt, W.J., 2002. Veterinary Parasitology Reference Manual. Wiley-Blackwell, Ames.

Fredes, F., Gorman, T., Silva, M., Alcaino, H., 1997. Evaluacion diagnostica de fracciones cromatograficas de *Fasciola hepatica* mediante Western Blot y ELISA en animales infectados. Arch. Med. Vet. 29, 283–294.

Gaudier, J.F., Caban-Hernandez, K., Osuna, A., Espino, A.M., 2012. Biochemical characterization and differential expression of a 16.5-kilodalton tegument-associated antigen from the liver fluke *Fasciola hepatica*. Clin. Vaccine Immunol. 19, 325–333.

Godoy, M.Y., Roque, L.E., Domenech, C.I., Rodriguez, F.R., 2010. Coproparasitological diagnosis of *Fasciola hepatica* in cattle in a Cuban livestock farm. Rev. Investig. Vet. Peru 21, 175–179.

Gonzalez-Lanza, C., Manga-Gonzalez, Y., Del-Pozo-Carnero, P., Hidalgo-Argüello, R., 1989. Dynamics of elimination of the eggs of *Fasciola hepatica* (Trematoda, Digenea) in the faeces of cattle in the Porma Basin, Spain. Vet. Parasitol. 34, 35–43.

Gordon, D.K., Zadoks, R.N., Stevenson, H., Sargison, N.D., Skuce, P.J., 2012. On farm evaluation of the coproantigen ELISA and coproantigen reduction test in Scottish sheep naturally infected with *Fasciola hepatica*. Vet. Parasitol. 187, 436–444.

Gorman, T., Aballay, J., Fredes, F., Silva, M., Aguillon, J.C., Alcaino, H.A., 1997. Immunodiagnosis of fasciolosis in horses and pigs using western blots. Int. J. Parasitol. 27, 1429–1432.

Gundlach, J.L., 1971. A study of the phenomena of immunity in the course of experimental fascioliasis in rabbits. Acta Parasitol. 19, 285–306.

Hammami, H., Ayadi, A., Camus, D., Dutoit, E., 1997. Diagnostic value of the demonstration of specific antigens of *Fasciola hepatica* by western blot technique. Parasite 4, 291–295.

Hammouda, N.A., El Mansoury, S.T., El Azzouni, M.Z., Hussein, E.D., 1997. Detection of circulating antigens in blood to evaluate treatment of fascioliasis. J. Egypt. Soc. Parasitol. 27, 365–371.

Happich, F.A., Boray, J.C., 1969a. Quantitative diagnosis of chronic fasciolosis. 1. Comparative studies on quantitative faecal examinations for chronic *Fasciola hepatica* infection in sheep. Aust. Vet. J. 45, 326–328.

Happich, F.A., Boray, J.C., 1969b. Quantitative diagnosis of chronic fasciolosis. 2. The estimation of daily total egg production of *Fasciola hepatica* and the number of adult flukes in sheep by faecal egg counts. Aust. Vet. J. 45, 329–331.

Haroun, E.T., Hillyer, G.V., 1986. Resistance to fascioliasis—a review. Vet. Parasitol. 20, 63–93.

Haroun, E.M., Elsanhouri, A.A., Gameel, A.A., 1989. Response of goats to repeated infections with Fasciola gigantica. Vet. Parasitol. 30, 287–296.

Haseeb, A.N., el-Shazly, A.M., Arafa, M.A., Morsy, A.T., 2003. Evaluation of excretory/secretory *Fasciola* (Fhes) antigen in diagnosis of human fascioliasis. J. Egypt. Soc. Parasitol. 33, 123–138.

Heussler, V.T., Dobbelaere, D.A.E., 1994. Cloning of a protease gene family of *Fasciola hepatica* by the polymerase chain-reaction. Mol. Biochem. Parasitol. 64, 11–23.

Heussler, V., Kaufmann, H., Strahm, D., Liz, J., Dobbelaere, D., 1993. DNA probes for the detection of *Fasciola hepatica* in snails. Mol. Cell. Probes 7, 261–267.

Hillyer, G.V., 1980. Isolation of *Fasciola hepatica* tegument antigens. J. Clin. Microbiol. 12, 695–699.

Hillyer, G.V., 1998. Immunodiagnosis of human and animal fasciolosis. In: Dalton, J.P. (Ed.), Fasciolosis. CABI Publishing, Oxon, Wallingford, UK, pp. 435–448.

Hillyer, G.V., Allain, D., 1979. Use of immunologic techniques to detect chemotherapeutic success in infections with *Fasciola hepatica*. III. Comparison of counter-electrophoresis and indirect hemagglutination in infected rabbits. J. Parasitol. 65, 960–963.

Hillyer, G.V., Cervoni, M., 1978. Laurell crossed immunoelectrophoresis and affinity chromatography for the purification of a parasite antigen. J. Immunol. Methods 20, 385–390.

Hillyer, G.V., Serrano, A.E., 1983. The antigens of *Paragonimus westermani*, *Schistosoma mansoni*, and *Fasciola hepatica* adult worms. Evidence for the presence of cross-reactive antigens and for cross-protection to *Schistosoma mansoni* infection using antigens of *Paragonimus westermani*. Am. J. Trop. Med. Hyg. 32, 350–358.

Hillyer, G.V., Serrano, A.E., 1986. Fractionation of *Fasciola hepatica* tegument antigens and their application to the serodiagnosis of experimental fascioliasis by the enzyme-linked immunosorbent assay. J. Helminthol. 60, 173–178.

Hillyer, G.V., Soler de Galanes, M., 1988. Identification of a 17-kilodalton *Fasciola hepatica* immunodiagnostic antigen by the enzyme-linked immunoelectrotransfer blot technique. J. Clin. Microbiol. 26, 2048–2053.

Hillyer, G.V., Soler de Galanes, M., 1991. Initial feasibility studies of the fast-ELISA for the immunodiagnosis of fascioliasis. J. Parasitol. 77, 362–365.

Hillyer, G.V., Taylor, D.W., 1988. Immunoprecipitation of *Fasciola hepatica* mRNA in vitro translation products using infection and hyperimmune sera. Am. J. Trop. Med. Hyg. 38, 547–552.

Hillyer, G.V., Soler de Galanes, M., Rodriguez-Perez, J., Bjorland, J., Silva de Lagrava, M., Ramirez Guzman, S., Bryan, R.T., 1992. Use of the Falcon assay screening test—enzyme-linked immunosorbent assay (FAST-ELISA) and the enzyme-linked immunoelectrotransfer blot (EITB) to determine the prevalence of human fascioliasis in the Bolivian Altiplano. Am. J. Trop. Med. Hyg. 46, 603–609.

Hodgkinson, J., Cwiklinski, K., Beesley, N.J., Paterson, S., Williams, D.J.L., 2013. Identification of putative markers of triclabendazole resistance by a genome-wide analysis of genetically recombinant *Fasciola hepatica*. Parasitology 140, 1523–1533.

Honer, M.R., 1965a. The interpretation of faecal egg-counts. I. Daily variations in Fasciola hepatica egg-counts in cattle. Z. Parasitenkd. 26, 143–155.

Honer, M.R., 1965b. The interpretation of faecal egg-counts. II. Single and multiple sampling in the diagnosis of sub-clinical fascioliasis hepatica. Z. Parasitenkd. 26, 156–162.

Honer, M.R., 1965c. The interpretation of faecal egg-counts. III. The influence of the age of the host on Fasciola hepatica egg-counts in cattle. Z. Parasitenkd. 26, 221–229.

Honer, M.R., 1967. The interpretation of faecal egg-counts. IV. The influence of faecal consistency and dry matter content on Fasciola hepatica egg-counts in cattle. Z. Parasitenkd. 28, 211–218.

Huang, C.J., Lin, H., Yang, X., 2012. Industrial production of recombinant therapeutics in Escherichia coli and its recent advancements. J. Ind. Microbiol. Biotechnol. 39, 383–399.

Hungerford, T.G., 1990. Diseases of Livestock. McGraw-Hill, Sydney.

Hurtrez-BoussÈS, S., Durand, P., Jabbour-Zahab, R., GuÉGan, J.F., Meunier, C., Bargues, M.D., Mas-Coma, S., Renaud, F., 2004. Isolation and characterization of microsatellite markers in the liver fluke (*Fasciola hepatica*). Mol. Ecol. Notes 4, 689–690.

Ibarra, F., Montenegro, N., Vera, Y., Boulard, C., Quiroz, H., Flores, J., Ochoa, P., 1998. Comparison of three ELISA tests for seroepidemiology of bovine fascioliosis. Vet. Parasitol. 77, 229–236.

Ichikawa, M., Itagaki, T., 2010. Discrimination of the ITS1 types of *Fasciola* spp. based on a PCR-RFLP method. Parasitol. Res. 106, 757–761.

Ikeda, T., 1998. Cystatin capture enzyme-linked immunosorbent assay for immunodiagnosis of human paragonimiasis and fascioliasis. Am. J. Trop. Med. Hyg. 59, 286–290.

Irving, D.O., Howell, M.J., 1982. Characterization of excretory-secretory antigens of *Fasciola hepatica*. Parasitology 85 (Pt 1), 179–188.

Itagaki, T., Sakaguchi, K., Terasaki, K., Sasaki, O., Yoshihara, S., Van Dung, T., 2009. Occurrence of spermic diploid and aspermic triploid forms of *Fasciola* in Vietnam and their molecular characterization based on nuclear and mitochondrial DNA. Parasitol. Int. 58, 81–85.

Kajugu, P.E., Hanna, R.E., Edgar, H.W., Forster, F.I., Malone, F.E., Brennan, G.P., Fairweather, I., 2012. Specificity of a coproantigen ELISA test for fasciolosis: lack of cross-reactivity with *Paramphistomum cervi* and *Taenia hydatigena*. Vet. Rec. 171, 502.

Karanis, P., Thekisoe, O., Kiouptsi, K., Ongerth, J., Igarashi, I., Inoue, N., 2007. Development and preliminary evaluation of a loop-mediated isothermal amplification procedure for sensitive detection of cryptosporidium oocysts in fecal and water samples. Appl. Environ. Microbiol. 73, 5660–5662.

Khalil, H.M., Abdel, T.M., Maklad, M.K., Abdallah, H.M., Fahmy, I.A., el Zayyat, E.A., 1990. Specificity of crude and purified *Fasciola* antigens in immunodiagnosis of human fascioliasis. J. Egypt. Soc. Parasitol. 20, 87–94.

Kim, K., Yang, H.J., Chung, Y.B., 2003. Usefulness of 8 kDa protein of *Fasciola hepatica* in diagnosis of fascioliasis. Korean J. Parasitol. 41, 121–123.

Klimenko, V.V., 1980. The occurrence of a specific *Fasciola hepatica* antigen in bile, a diagnostic criterion in fascioliasis. Parazitologiya 14, 441–451.

Knobloch, J., 1985. Human fascioliasis in Cajamarca/Peru. II. Humoral antibody response and antigenaemia. Trop. Med. Parasitol. 36, 91–93.

Koboldt, D.C., Steinberg, K.M., Larson, D.E., Wilson, R.K., Mardis, E.R., 2013. The next-generation sequencing revolution and its impact on genomics. Cell 155, 27–38.

Korach, S., Benex, J., 1966. A lipoprotein antigen in *Fasciola hepatica*. II. Immunological and immunochemical properties. Exp. Parasitol. 19, 199–205.

Kozak, M., Garbacewicz, A., Wędrychowicz, H., 2008. The performance of a PCR assay for the detection of *Fasciola hepatica* DNA in fecal samples. Wiad. Parazytol. 54.

Kuboki, N., Inoue, N., Sakurai, T., Di Cello, F., Grab, D.J., Suzuki, H., Sugimoto, C., Igarashi, I., 2003. Loop-mediated isothermal amplification for detection of African trypanosomes. J. Clin. Microbiol. 41, 5517–5524.

Kuerpick, B., Schnieder, T., Strube, C., 2012. Seasonal pattern of *Fasciola hepatica* antibodies in dairy herds in Northern Germany. Parasitol. Res. 111, 1085–1092.

Kuerpick, B., Schnieder, T., Strube, C., 2013. Evaluation of a recombinant cathepsin L1 ELISA and comparison with the Pourquier and ES ELISA for the detection of antibodies against *Fasciola hepatica*. Vet. Parasitol. 193, 206–213.

Lammas, D.A., Duffus, W.P., 1983. The shedding of the outer glycocalyx of juvenile *Fasciola hepatica*. Vet. Parasitol. 12, 165–178.

Langley, R.J., Hillyer, G.V., 1989. Detection of circulating parasite antigen in murine fascioliasis by two-site enzyme-linked immunosorbent assays. Am. J. Trop. Med. Hyg. 41, 472–478.

Lavier, G., Stefanopoulo, G., 1944. L'intradermo-reaction et la reaction de fixation du complement dans la distomatqpe humaine a *Fasciola hepatica*. Bull. Soc. Pathol. Exot. 37, 302–310.

Le, T.H., De, N.V., Agatsuma, T., Thi Nguyen, T.G., Nguyen, Q.D., McManus, D.P., Blair, D., 2008. Human fascioliasis and the presence of hybrid/introgressed forms of *Fasciola hepatica* and *Fasciola gigantica* in Vietnam. Int. J. Parasitol. 38, 725–730.

Le, T.H., Nguyen, K.T., Nguyen, N.T., Doan, H.T., Le, X.T., Hoang, C.T., De, N.V., 2012a. Development and evaluation of a single-step duplex PCR for simultaneous detection of *Fasciola hepatica* and *Fasciola gigantica* (family Fasciolidae, class Trematoda, phylum Platyhelminthes). J. Clin. Microbiol. 50, 2720–2726.

Le, T.H., Nguyen, N.T., Truong, N.H., De, N.V., 2012b. Development of mitochondrial loop-mediated isothermal amplification for detection of the small liver fluke *Opisthorchis viverrini* (Opisthorchiidae; Trematoda; Platyhelminthes). J. Clin. Microbiol. 50, 1178–1184.

Leclipteux, T., Torgerson, P.R., Doherty, M.L., McCole, D., Protz, M., Farnir, F., Losson, B., 1998. Use of excretory/secretory antigens in a competition test to follow the kinetics of infection by *Fasciola hepatica* in cattle. Vet. Parasitol. 77, 103–114.

Lehner, R.P., Sewell, M.M.H., 1980. A study of the antigens produced by adult *Fasciola hepatica* maintained in vitro. Parasite Immunol. 2, 99–109.

Levieux, D., Levieux, A., 1994. Early immunodiagnosis of caprine fasciolosis using the specific f2 antigen in a passive hemagglutination test. Vet. Parasitol. 53, 59–66.

Levieux, D., Levieux, A., Mage, C., Venien, A., 1992a. Early immunodiagnosis of bovine fascioliasis using the specific antigen f2 in a passive hemagglutination test. Vet. Parasitol. 44, 77–86.

Levieux, D., Levieux, A., Venien, A., 1992b. An improved passive hemagglutination test for the serological diagnosis of bovine fascioliasis using the specific antigen f2. Vet. Parasitol. 42, 53–66.

Li, O., Leguía, G., Espino, A.M., Duménigo, B., Díaz, A., Otero, O., 2005. Detección de anticuerpos y antígenos para el diagnóstico de *Fasciola hepatica* en alpacas naturalmente infectadas. Rev. Investig. Vet. Peru 16, 143–153.

Lotfollahzadeh, S., Mohri, M., Bahadori Sh, R., Dezfouly, M.R., Tajik, P., 2008. The relationship between normocytic, hypochromic anaemia and iron concentration together with hepatic enzyme activities in cattle infected with *Fasciola hepatica*. J. Helminthol. 82, 85–88.

Maes, L., Lauwers, H., Deckers, W., Vanparijs, O., 1988. Flukicidal action of closantel against immature and mature *Fasciola hepatica* in experimentally infected rats and sheep. Res. Vet. Sci. 44, 229–232.

Magalhaes, K.G., Passos, L.K.J., Carvalho, O.dS., 2004. Detection of *Lymnaea columella* infection by *Fasciola hepatica* through multiplex-PCR. Mem. Inst. Oswaldo Cruz 99, 421–424.

Magalhaes, K.G., Jannotti-Passos, L.K., Caldeira, R.L., Berne, M.E.A., Muller, G., Carvalho, O.S., Lenzi, H.L., 2008. Isolation and detection of *Fasciola hepatica* DNA in *Lymnaea viatrix* from formalin-fixed and paraffin-embedded tissues through multiplex-PCR. Vet. Parasitol. 152, 333–338.

Maher, K., El Ridi, R., Elhoda, A.N., El-Ghannam, M., Shaheen, H., Shaker, Z., Hassanein, H.I., 1999. Parasite-specific antibody profile in human fascioliasis: application for immunodiagnosis of infection. Am. J. Trop. Med. Hyg. 61, 738–742.

Marcet Sanchez, R., Figueredo Pino, M., Nunez Fernandez, C.F., Rojas Rivero, C.L., Sarracent Perez, C.J., 2012. Increase of analytical sensitivity of FasciDIG system for the diagnosis of Fasciola hepatica. Rev. Cubana Med. Trop. 64, 335–341.

Marcilla, A., Bargues, M.D., Mas-Coma, S., 2002. A PCR-RFLP assay for the distinction between *Fasciola hepatica* and *Fasciola gigantica*. Mol. Cell. Probes 16, 327–333.

Marcilla, A., De la Rubia, J.E., Sotillo, J., Bernal, D., Carmona, C., Villavicencio, Z., Acosta, D., Tort, J., Bornay, F.J., Esteban, J.G., Toledo, R., 2008. Leucine aminopeptidase is an immunodominant antigen of *Fasciola hepatica* excretory and secretory products in human infections. Clin. Vaccine Immunol. 15, 95–100.

Martinez, A., Martinez-Cruz, M.S., Martinez, F.J., Gutierrez, P.N., Hernandez, S., 1996. Detection of antibodies of Fasciola hepatica excretory–secretory antigens in experimentally infected goats by enzyme immunosorbent assay. Vet. Parasitol. 62, 247–252.

Martinez-Ibeas, A.M., Gonzalez-Warleta, M., Martinez-Valladares, M., Castro-Hermida, J.A., Gonzalez-Lanza, C., Minambres, B., Ferreras, C., Mezo, M., Manga-Gonzalez, M.Y., 2013. Development and validation of a mtDNA multiplex PCR for identification and discrimination of *Calicophoron daubneyi* and *Fasciola hepatica* in the *Galba truncatula* snail. Vet. Parasitol. 195, 57–61.

Martinez-Moreno, A., Martinez-Moreno, F.J., Acosta, I., Gutierrez, P.N., Becerra, C., Hernandez, S., 1997. Humoral and cellular immune responses to experimental *Fasciola hepatica* infections in goats. Parasitol. Res. 83, 680–686.

Martinez-Perez, J.M., Robles-Perez, D., Rojo-Vazquez, F.A., Martinez-Valladares, M., 2012. Comparison of three different techniques to diagnose *Fasciola hepatica* infection in experimentally and naturally infected sheep. Vet. Parasitol. 190, 80–86.

Martinez-Sernandez, V., Muino, L., Perteguer, M.J., Garate, T., Mezo, M., Gonzalez-Warleta, M., Muro, A., Correia da Costa, J.M., Romaris, F., Ubeira, F.M., 2011. Development and evaluation of a new lateral flow immunoassay for serodiagnosis of human fasciolosis. PLoS Negl. Trop. Dis. 5, e1376.

Martinez-Valladares, M., Cordero-Perez, C., Rojo-Vazquez, F.A., 2014. Efficacy of an anthelmintic combination in sheep infected with Fasciola hepatica resistant to albendazole and clorsulon. Exp. Parasitol. 136, 59–62.

Mas-Coma, S., 2005. Epidemiology of fascioliasis in human endemic areas. J. Helminthol. 79, 207–216.

Mas-Coma, S., Bargues, M.D., Esteban, J.G., 1999. Human fasciolosis. In: Dalton, J.P. (Ed.), Fasciolosis. CABI Publishing, Oxon, UK, pp. 411–434.

Mas-Coma, S., Valero, M.A., Bargues, M.D., 2009. Chapter 2. *Fasciola*, lymnaeids and human fascioliasis, with a global overview on disease transmission, epidemiology, evolutionary genetics, molecular epidemiology and control. Adv. Parasitol. 69, 41–146.

McCaughey, W.J., Hatch, C., 1964. Routine faecal examination for the detection of fluke (*Fasciola hepatica*) eggs: an evaluation of some techniques. Ir. Vet. J. 18, 181–187.

McGarry, J.W., Ortiz, P.L., Hodgkinson, J.E., Goreish, I., Williams, D.J., 2007. PCR-based differentiation of *Fasciola* species (Trematoda: Fasciolidae), using primers based on RAPD-derived sequences. Ann. Trop. Med. Parasitol. 101, 415–421.

McManus, D.P., 2014. Immunodiagnosis of sheep infections with *Echinococcus granulosus*: in 35 years where have we come? Parasite Immunol. 36, 125–130.

Mejia, R., Vicuna, Y., Broncano, N., Sandoval, C., Vaca, M., Chico, M., Cooper, P.J., Nutman, T.B., 2013. A novel, multi-parallel, real-time polymerase chain reaction approach for eight gastrointestinal parasites provides improved diagnostic capabilities to resource-limited at-risk populations. Am. J. Trop. Med. Hyg. 88, 1041–1047.

Menard, A., L'Hostis, M., Leray, G., Marchandeau, S., Pascal, M., Roudot, N., Michel, V., Chauvin, A., 2000. Inventory of wild rodents and lagomorphs as natural hosts of Fasciola hepatica on a farm located in a humid area in Loire Atlantique (France). Parasite 7, 77–82.

Mezo, M., Gonzalez-Warleta, M., Ubeira, F.M., 2003. Optimized serodiagnosis of sheep fascioliasis by fast-D protein liquid chromatography fractionation of *Fasciola hepatica* excretory–secretory antigens. J. Parasitol. 89, 843–849.

Mezo, M., Gonzalez-Warleta, M., Carro, C., Ubeira, F.M., 2004. An ultrasensitive capture ELISA for detection of *Fasciola hepatica* coproantigens in sheep and cattle using a new monoclonal antibody (MM3). J. Parasitol. 90, 845–852.

Mezo, M., Gonzalez-Warleta, M., Ubeira, F.M., 2007. The use of MM3 monoclonal antibodies for the early immunodiagnosis of ovine fascioliasis. J. Parasitol. 93, 65–72.

Mezo, M., Gonzalez-Warleta, M., Castro-Hermida, J.A., Carro, C., Ubeira, F.M., 2010a. Kinetics of anti-*Fasciola* IgG antibodies in serum and milk from dairy cows during lactation, and in serum from calves after feeding colostrum from infected dams. Vet. Parasitol. 168, 36–44.

Mezo, M., Gonzalez-Warleta, M., Castro-Hermida, J.A., Muino, L., Ubeira, F.M., 2010b. Field evaluation of the MM3-SERO ELISA for detection of anti-*Fasciola* IgG antibodies in milk samples from individual cows and bulk milk tanks. Parasitol. Int. 59, 610–615.

Mezo, M., Gonzalez-Warleta, M., Castro-Hermida, J.A., Muino, L., Ubeira, F.M., 2011. Association between anti-*F. hepatica* antibody levels in milk and production losses in dairy cows. Vet. Parasitol. 180, 237–242.

Moll, L., Gaasenbeek, C.P., Vellema, P., Borgsteede, F.H., 2000. Resistance of *Fasciola hepatica* against triclabendazole in cattle and sheep in The Netherlands. Vet. Parasitol. 91, 153–158.

Molloy, J.B., Anderson, G.R., Fletcher, T.I., Landmann, J., Knight, B.C., 2005. Evaluation of a commercially available enzyme-linked immunosorbent assay for detecting antibodies to *Fasciola hepatica* and *Fasciola gigantica* in cattle, sheep and buffaloes in Australia. Vet. Parasitol. 130, 207–212.

Morales, A., Espino, A.M., 2012. Evaluation and characterization of *Fasciola hepatica* tegument protein extract for serodiagnosis of human fascioliasis. Clin. Vaccine Immunol. 19, 1870–1878.

Moustafa, N.E., Hegab, M.H., Hassan, M.M., 1998. Role of ELISA in early detection of *Fasciola* copro-antigens in experimentally infected animals. J. Egypt. Soc. Parasitol. 28, 379–387.

Moxon, J.V., LaCourse, E.J., Wright, H.A., Perally, S., Prescott, M.C., Gillard, J.L., Barrett, J., Hamilton, J.V., Brophy, P.M., 2010. Proteomic analysis of embryonic *Fasciola hepatica*: characterization and antigenic potential of a developmentally regulated heat shock protein. Vet. Parasitol. 169, 62–75.

Muino, L., Perteguer, M.J., Garate, T., Martinez-Sernandez, V., Beltran, A., Romaris, F., Mezo, M., Gonzalez-Warleta, M., Ubeira, F.M., 2011. Molecular and immunological characterization of *Fasciola* antigens recognized by the MM3 monoclonal antibody. Mol. Biochem. Parasitol. 179, 80–90.

Mulcahy, G., Joyce, P., Dalton, J.P., 1999. Immunology of *Fasciola hepatica* infection. In: Dalton, J.P. (Ed.), Fasciolosis. CAB International, Oxon, United Kingdom, pp. 341–375.

Muro, A., Martin, F.S., Rodriguez-Medina, J.R., Hillyer, G.V., 1994. Identification of genes that encode *Fasciola*-specific arc 2 antigens. Am. J. Trop. Med. Hyg. 51, 684–689.

Neyra, V., Chavarry, E., Espinoza, J.R., 2002. Cysteine proteinases Fas1 and Fas2 are diagnostic markers for *Fasciola hepatica* infection in alpacas (Lama pacos). Vet. Parasitol. 105, 21–32.

Njiru, Z.K., Mikosza, A.S., Armstrong, T., Enyaru, J.C., Ndung'u, J.M., Thompson, A.R., 2008. Loop-mediated isothermal amplification (LAMP) method for rapid detection of *Trypanosoma brucei rhodesiense*. PLoS Negl. Trop. Dis. 2, e147.

Novobilsky, A., Averpil, H.B., Hoglund, J., 2012. The field evaluation of albendazole and triclabendazole efficacy against *Fasciola hepatica* by coproantigen ELISA in naturally infected sheep. Vet. Parasitol. 190, 272–276.

O'Neill, S.M., Parkinson, M., Strauss, W., Angles, R., Dalton, J.P., 1998. Immunodiagnosis of *Fasciola hepatica* infection (fascioliasis) in a human population in the Bolivian Altiplano using purified cathepsin L cysteine proteinase. Am. J. Trop. Med. Hyg. 58, 417–423.

O'Neill, S.M., Parkinson, M., Dowd, A.J., Strauss, W., Angles, R., Dalton, J.P., 1999. Short report: immunodiagnosis of human fascioliasis using recombinant *Fasciola hepatica* cathepsin L1 cysteine proteinase. Am. J. Trop. Med. Hyg. 60, 749–751.

Ortiz, P.L., Claxton, J.R., Clarkson, M.J., McGarry, J., Williams, D.J.L., 2000. The specificity of antibody responses in cattle naturally exposed to *Fasciola hepatica*. Vet. Parasitol. 93, 121–134.

Ortiz, P., Scarcella, S., Cerna, C., Rosales, C., Cabrera, M., Guzman, M., Lamenza, P., Solana, H., 2013. Resistance of *Fasciola hepatica* against Triclabendazole in cattle in Cajamarca (Peru): a clinical trial and an in vivo efficacy test in sheep. Vet. Parasitol. 195, 118–121.

Overend, D.J., Bowen, F.L., 1995. Resistance of *Fasciola hepatica* to triclabendazole. Aust. Vet. J. 72, 275–276.

Paz-Silva, A., Pedreira, J., Sanchez-Andrade, R., Suarez, J.L., Diaz, P., Panadero, R., Diez-Banos, P., Morrondo, P., 2002. Time-course analysis of coproantigens in rats infected and challenged with *Fasciola hepatica*. Parasitol. Res. 88, 568–573.

Paz-Silva, A., Sanchez-Andrade, R., Suarez, J.L., Pedreira, J., Arias, M., Lopez, C., Panadero, R., Diaz, P., Diez-Banos, P., Morrondo, P., 2003. Prevalence of natural ovine fasciolosis shown by demonstrating the presence of serum circulating antigens. Parasitol. Res. 91, 328–331.

Paz-Silva, A., Hillyer, G.V., Sanchez-Andrade, R., Rodriguez-Medina, J.R., Arias, M., Morrondo, P., Diez-Banos, P., 2005. Isolation, identification and expression of a *Fasciola hepatica* cDNA encoding a 2.9-kDa recombinant protein for the diagnosis of ovine fasciolosis. Parasitol. Res. 95, 129–135.

Pelayo, L., Espino, A.M., Dumenigo Ripoll, B.E., Finlay Villalvilla, C.M., 1998. The detection of antibodies, antigens and circulating immune complexes in acute and chronic fascioliasis. Preliminary results. Rev. Cubana Med. Trop. 50, 209–214.

Peng, M., Ichinomiya, M., Ohtori, M., Ichikawa, M., Shibahara, T., Itagaki, T., 2009. Molecular characterization of *Fasciola hepatica*, *Fasciola gigantica*, and aspermic *Fasciola* sp. in China based on nuclear and mitochondrial DNA. Parasitol. Res. 105, 809–815.

Pfister, K., Daveau, C., Ambroise-Thomas, P., 1984. Partial purification of somatic and excretory-secretory products of adult *Fasciola hepatica* and their application for the serodiagnosis of experimental and natural fascioliasis using an ELISA. Res. Vet. Sci. 37, 39–43.

Piacenza, L., Acosta, D., Dowd, A., McGonicle, S., Dalton, J., Carmona, C., 1997. Proteinases secreted by *Fasciola hepatica*: time course of the inhibitory effect of serum from experimentally infected rabbits demonstrated by gelatin-substrate polyacrylamide gel electrophoresis. J. Helminthol. 71, 333–338.

Piedrafita, D., Raadsma, H.W., Prowse, R., Spithill, T.W., 2004. Immunology of the host–parasite relationship in fasciolosis (*Fasciola hepatica* and *Fasciola gigantica*). Can. J. Zool. 82, 233–250.

Qureshi, T., Wagner, G.G., Drawe, D.L., Davis, D.S., Craig, T.M., 1995. Enzyme-linked immunoelectrotransfer blot analysis of excretory-secretory proteins of *Fascioloides magna* and *Fasciola hepatica*. Vet. Parasitol. 58, 357–363.

Rapsch, C., Schweizer, G., Grimm, F., Kohler, L., Bauer, C., Deplazes, P., Braun, U., Torgerson, P.R., 2006. Estimating the true prevalence of *Fasciola hepatica* in cattle slaughtered in Switzerland in the absence of an absolute diagnostic test. Int. J. Parasitol. 36, 1153–1158.

Reichel, M.P., 2002. Performance characteristics of an enzyme-linked immunosorbent assay for the detection of liver fluke (*Fasciola hepatica*) infection in sheep and cattle. Vet. Parasitol. 107, 65–72.

Reichel, M.P., Vanhoff, K., Baxter, B., 2005. Performance characteristics of an enzyme-linked immunosorbent assay performed in milk for the detection of liver fluke (*Fasciola hepatica*) infection in cattle. Vet. Parasitol. 129, 61–66.

Rickard, L.G., 1995. Development and application of a dot-ELISA test for the detection of serum antibodies to *Fasciola hepatica* antigens in llamas. Vet. Parasitol. 58, 9–15.

Rivera Marrero, C.A., Santiago, N., Hillyer, G.V., 1988. Evaluation of immunodiagnostic antigens in the excretory–secretory products of *Fasciola hepatica*. J. Parasitol. 74, 646–652.

Robert, R., Jarrige, P.L.d.l., Chabasse, D., Mahaza, C., Bizon, C., Genthon, H., 1980. Immunological diagnosis of *Fasciola hepatica* infection in cattle. Detection of antigen fraction II and antibody antifraction II. Rec. Med. Vet. Ec. Alfort 156, 533–538.

Roberts, J.A., Estuningsih, E., Widjayanti, S., Wiedosari, E., Partoutomo, S., Spithill, T.W., 1997. Resistance of Indonesian thin tail sheep against *Fasciola gigantica* and *F. hepatica*. Vet. Parasitol. 68, 69–78.

Robinson, M.W., Dalton, J.P., Donnelly, S., 2008. Helminth pathogen cathepsin proteases: it's a family affair. Trends Biochem. Sci. 33, 601–608.

Robles-Perez, D., Martinez-Perez, J.M., Rojo-Vazquez, F.A., Martinez-Valladares, M., 2013. The diagnosis of fasciolosis in feces of sheep by means of a PCR and its application in the detection of anthelmintic resistance in sheep flocks naturally infected. Vet. Parasitol. 197, 277–282.

Rodriguez-Perez, J., Hillyer, G.V., 1995. Detection of excretory-secretory circulating antigens in sheep infected with *Fasciola hepatica* and with *Schistosoma mansoni* and *F. hepatica*. Vet. Parasitol. 56, 57–66.

Roeber, F., Larsen, J.W., Anderson, N., Campbell, A.J., Anderson, G.A., Gasser, R.B., Jex, A.R., 2012. A molecular diagnostic tool to replace larval culture in conventional faecal egg count reduction testing in sheep. PLoS One 7, e37327.

Rognlie, M.C., Dimke, K.L., Knapp, S.E., 1994. Detection of *Fasciola hepatica* in infected intermediate hosts using RT-PCR. J. Parasitol. 80, 748–755.

Rokni, M.B., Massoud, J., O'Neill, S.M., Parkinson, M., Dalton, J.P., 2002. Diagnosis of human fasciolosis in the Gilan province of Northern Iran: application of cathepsin L-ELISA. Diagn. Microbiol. Infect. Dis. 44, 175–179.

Rokni, M.B., Mirhendi, H., Mizani, A., Mohebali, M., Sharbatkhori, M., Kia, E.B., Abdoli, H., Izadi, S., 2010. Identification and differentiation of *Fasciola hepatica* and

Fasciola gigantica using a simple PCR-restriction enzyme method. Exp. Parasitol. 124, 209–213.

Romasanta, A., Romero, J.L., Arias, M., Sanchez-Andrade, R., Lopez, C., Suarez, J.L., Diaz, P., Diez-Banos, P., Morrondo, P., Paz-Silva, A., 2003. Diagnosis of parasitic zoonoses by immunoenzymatic assays—analysis of cross-reactivity among the excretory/secretory antigens of *Fasciola hepatica*, *Toxocara canis*, and *Ascaris suum*. Immunol. Invest. 32, 131–142.

Ruiz, A., Molina, J.M., Gonzalez, J., Martinez-Moreno, F.J., Gutierrez, P.N., Martinez-Moreno, A., 2003. Humoral response (IgG) of goats experimentally infected with *Fasciola hepatica* against cysteine proteinases of adult fluke. Vet. Res. 34, 435–443.

Salimi-Bejestani, M.R., Daniel, R.G., Felstead, S.M., Cripps, P.J., Mahmoody, H., Williams, D.J., 2005. Prevalence of *Fasciola hepatica* in dairy herds in England and Wales measured with an ELISA applied to bulk-tank milk. Vet. Rec. 156, 729–731.

Salimi-Bejestani, M.R., Daniel, R., Cripps, P., Felstead, S., Williams, D.J.L., 2007. Evaluation of an enzyme-linked immunosorbent assay for detection of antibodies to *Fasciola hepatica* in milk. Vet. Parasitol. 149, 290–293.

Sampaio-Silva, M.L., Da Costa, J.M., Da Costa, A.M., Pires, M.A., Lopes, S.A., Castro, A.M., Monjour, L., 1996. Antigenic components of excretory-secretory products of adult *Fasciola hepatica* recognized in human infections. Am. J. Trop. Med. Hyg. 54, 146–148.

Sanchez-Andrade, R., Paz-Silva, A., Suarez, J., Panadero, R., Diez-Banos, P., Morrondo, P., 2000. Use of a sandwich-enzyme-linked immunosorbent assay (SEA) for the diagnosis of natural *Fasciola hepatica* infection in cattle from Galicia (NW Spain). Vet. Parasitol. 93, 39–46.

Sanchez-Andrade, R., Paz-Silva, A., Suarez, J.L., Panadero, R., Pedreira, J., Diez-Banos, P., Morrondo, P., 2001. Effect of fasciolicides on the antigenaemia in sheep naturally infected with *Fasciola hepatica*. Parasitol. Res. 87, 609–614.

Sandeman, R.M., Howell, M.J., 1981. Precipitating antibodies against excretory/secretory antigens of *Fasciola hepatica* in sheep serum. Vet. Parasitol. 9, 35–46.

Santiago de Weil, N., Hillyer, G.V., Pacheco, E., 1984. Isolation of *Fasciola hepatica* genus-specific antigens. Int. J. Parasitol. 14, 197–206.

Santiago, N., Hillyer, G.V., 1988. Antibody profiles by EITB and ELISA of cattle and sheep infected with *Fasciola hepatica*. J. Parasitol. 74, 810–818.

Santiago, N., Hillyer, G.V., Garcia-Rosa, M., Morales, M.H., 1986. Identification of functional *Fasciola hepatica* antigens in experimental infections in rabbits. Am. J. Trop. Med. Hyg. 35, 135–140.

Semyenova, S.K., Morozova, E.V., Chrisanfova, G.G., Asatrian, A.M., Movsessian, S.O., 2003. RAPD variability and genetic diversity in two populations of liver fluke, *Fasciola hepatica*. Acta Parasitol. 48, 125–130.

Sewell, M.M., Hammond, J.A., 1972. The detection of *Fasciola* eggs in faeces. Vet. Rec. 90, 510–511.

Sexton, J.L., Milner, A.R., Campbell, N.J., 1991. *Fasciola hepatica*: immunoprecipitation analysis of biosynthetically labelled antigens using sera from infected sheep. Parasite Immunol. 13, 105–108.

Sievers, H.K., Oyarzun, R., 1932. Diagnostic de la discomatose hepatique par la reaction allergique. C. R. Seances Soc. Biol. 110, 630–632.

Silva, E., Castro, A., Lopes, A., Rodrigues, A., Dias, C., Conceicao, A., Alonso, J., Correia da Costa, J.M., Bastos, M., Parra, F., Moradas-Ferreira, P., Silva, M., 2004. A recombinant antigen recognized by *Fasciola hepatica*-infected hosts. J. Parasitol. 90, 746–751.

Soulsby, E.J., 1954. Skin hypersensitivity in cattle infested with *Fasciola hepatica*. J. Comp. Pathol. 64, 267–274.

Spithill, T., Smooker, P.M., Copeman, B., 1999. *Fasciola gigantica*: epidemiology, control, immunology and molecular biology. In: Dalto, J.P. (Ed.), Fasciolosis. CABI, Oxon, UK, pp. 465–525.

Stark, D., Al-Qassab, S.E., Barratt, J.L., Stanley, K., Roberts, T., Marriott, D., Harkness, J., Ellis, J.T., 2011. Evaluation of multiplex tandem real-time PCR for detection of *Cryptosporidium* spp., *Dientamoeba fragilis*, *Entamoeba histolytica*, and *Giardia intestinalis* in clinical stool samples. J. Clin. Microbiol. 49, 257–262.

Sykes, A.R., Coop, R.L., Robinson, M.G., 1980. Chronic subclinical ovine fascioliasis: plasma glutamate dehydrogenase, gamma-glutamyl transpeptidase and aspartate aminotransferase activities and their significance as diagnostic aids. Res. Vet. Sci. 28, 71–75.

Szaflarski, J., 1950. Zastosowanie proby allergicznej srodskorno-powiekowej w diagnostyce chorob pasozytniczych u zwierzqt. Med. Weter. 6, 585–589.

Tailliez, R., 1967. Specific antigen of *Fasciola hepatica* and its application to clinical diagnosis of fascioliasis. International Liverfluke Colloquium. Wageningen, Netherlands, pp. 172–173; 194–195.

Tailliez, R., Korach, S., 1970. *Fasciola hepatica* antigens. I. Isolation and characterization of a genus-specific antigen. Ann. Inst. Pasteur 118, 61–78.

Taniuchi, M., Verweij, J.J., Noor, Z., Sobuz, S.U., Lieshout, L., Petri Jr., W.A., Haque, R., Houpt, E.R., 2011. High throughput multiplex PCR and probe-based detection with Luminex beads for seven intestinal parasites. Am. J. Trop. Med. Hyg. 84, 332–337.

Teodorovic, D., Berkes, I., Milovanavic, M., 1963. Diagnosis of liver fluke (Fasciola hepatica) infection in human beings by means of immunoelectrophoresis. Nature 198, 204.

Thorpe, E., Ford, E.J.H., 1969. Serum enzyme and hepatic changes in sheep infested with *Fasciola hepatica*. J. Pathol. 97, 619–629.

Timoteo, O., Maco Jr., V., Maco, V., Neyra, V., Yi, P.J., Leguia, G., Espinoza, J.R., 2005. Characterization of the humoral immune response in alpacas (Lama pacos) experimentally infected with *Fasciola hepatica* against cysteine proteinases Fas1 and Fas2 and histopathological findings. Vet. Immunol. Immunopathol. 106, 77–86.

Torgerson, P., Claxton, J., 1999. Epidemiology and control. In: Dalton, J.P. (Ed.), Fasciolosis. CABI, Oxon, UK, pp. 113–149.

Trudgett, A., Anderson, A., Hanna, R.E., 1988. Use of immunosorbent-purified antigens of *Fasciola hepatica* in enzyme immunoassays. Res. Vet. Sci. 44, 262–263.

Ubeira, F.M., Muino, L., Valero, M.A., Periago, M.V., Perez-Crespo, I., Mezo, M., Gonzalez-Warleta, M., Romaris, F., Paniagua, E., Cortizo, S., Llovo, J., Mas-Coma, S., 2009. MM3-ELISA detection of *Fasciola hepatica* coproantigens in preserved human stool samples. Am. J. Trop. Med. Hyg. 81, 156–162.

Urquhart, G.M., Duncan, J., Armour, L., Dunn, J., Jennings, A.M., 1996. Veterinary Parasitology. Blackwell Science, UK.

Valero, M.A., Ubeira, F.M., Khoubbane, M., Artigas, P., Muino, L., Mezo, M., Perez-Crespo, I., Periago, M.V., Mas-Coma, S., 2009. MM3-ELISA evaluation of coproantigen release and serum antibody production in sheep experimentally infected with *Fasciola hepatica* and *F. gigantica*. Vet. Parasitol. 159, 77–81.

Valero, M.A., Periago, M.V., Perez-Crespo, I., Angles, R., Villegas, F., Aguirre, C., Strauss, W., Espinoza, J.R., Herrera, P., Terashima, A., Tamayo, H., Engels, D., Gabrielli, A.F., Mas-Coma, S., 2012a. Field evaluation of a coproantigen detection test for fascioliasis diagnosis and surveillance in human hyperendemic areas of Andean countries. PLoS Negl. Trop. Dis. 6, e1812.

Valero, M.A., Periago, M.V., Perez-Crespo, I., Rodriguez, E., Perteguer, M.J., Garate, T., Gonzalez-Barbera, E.M., Mas-Coma, S., 2012b. Assessing the validity of an ELISA test for the serological diagnosis of human fascioliasis in different epidemiological situations. Trop. Med. Int. Health 17, 630–636.

Vargas, D., Vega, M., Gonzalez, C.G., 2003. Aproximación a una caracterización molecular de *Fasciola hepatica* por la técnica RAPDs—PCR. Parasitologia Latinoamericana 58, 11–16.

Vercruysse, J., Claerebout, E., 2001. Treatment vs non-treatment of helminth infections in cattle: defining the threshold. Vet. Parasitol. 98, 195–214.

Wagner, O., 1935. Hautallergie und Komplementbiridungsreaktion bei Trernatodeninfektionen. Z. Immun. Exp. Ther. 84, 225–236.

Waseem, S., Khalid, M., Rashid, M., Waqar, A., Muhammad, I., Rashid, A., Khan, M.S., Sabir, A.J., 2012. Prevalence and molecular diagnosis of *Fasciola hepatica* in sheep and goats in different districts of Punjab, Pakistan. Pak. Vet. J. 32, 535–538.

Wuhrer, M., Grimm, C., Dennis, R.D., Idris, M.A., Geyer, R., 2004. The parasitic trematode *Fasciola hepatica* exhibits mammalian-type glycolipids as well as Gal(beta1-6)Gal-terminating glycolipids that account for cestode serological cross-reactivity. Glycobiology 14, 115–126.

Yamasaki, H., Aoki, T., Oya, H., 1989. A cysteine proteinase from the liver fluke *Fasciola* spp.: purification, characterization, localization and application to immunodiagnosis. Jpn. J. Parasitol. 38, 373–384.

Zimmerman, G.L., Clark, C.R.B., 1986. Separation of parasite antigens by molecular exclusion, anion exchange, and chromatofocusing utilizing FPLC protein fractionation systems. Vet. Parasitol. 20, 217–228.

Zimmerman, G.L., Jen, L.W., Cerro, J.E., Farnsworth, K.L., Wescott, R.B., 1982. Diagnosis of *Fasciola hepatica* infections in sheep by an enzyme-linked immunosorbent assay. Am. J. Vet. Res. 43, 2097–2100.

CHAPTER THREE

Reevaluating the Evidence for *Toxoplasma gondii*-Induced Behavioural Changes in Rodents

Amanda R. Worth[*,1], **R.C. Andrew Thompson**[*], **Alan J. Lymbery**[*,†]

[*]Parasitology, School of Veterinary and Life Sciences, Murdoch University, Perth, Western Australia, Australia
[†]Freshwater Fish Group & Fish Health Unit, School of Veterinary and Life Sciences, Murdoch University, Perth, Western Australia, Australia
[1]Corresponding author: e-mail address: a.worth@murdoch.edu.au

Contents

Advances in Parasitology, Volume 85
ISSN 0065-308X
http://dx.doi.org/10.1016/B978-0-12-800182-0.00003-9

Abstract

The ubiquitous protozoan parasite *Toxoplasma gondii* has been associated with behavioural changes in various hosts, including humans. In rodents, these behavioural changes are thought to represent adaptive manipulation by *T. gondii* to enhance transmission from intermediate hosts to the feline definitive host. In this review, we have tabulated evidence of changes in motor coordination, learning, memory, locomotion, anxiety, response to novelty and aversion to feline odour in rodents experimentally infected with *T. gondii*. In general, there was no consistent indication of the direction or magnitude of behavioural changes in response to infection. This may be due to the use, in these experimental studies, of different *T. gondii* strains, different host species and sexes and/or different methodologies to measure behaviour. A particular problem with studies of behavioural manipulation is likely to be the validity of behavioural tests, that is, whether they are actually measuring the traits that they were designed to measure.

We suggest that future studies can be improved in three major ways. First, they should use multiple tests of behaviour, followed by multivariate data analysis to identify behavioural constructs such as aversion, anxiety and response to novelty. Second, they should incorporate longitudinal measurements on the behaviour of individual hosts before and after infection, so that within-individual and between-individual variances and covariances in behavioural traits can be estimated. Finally, they should investigate how variables such as parasite strain, host species and host sex interact with parasite infection to alter host behaviour, in order to provide a sound foundation for research concerning the proximate and ultimate mechanism(s) responsible for behavioural changes.

1. INTRODUCTION

Toxoplasma gondii (Apicomplexa: Coccidia) is an intracellular parasite that infects warm-blooded animals, including humans, throughout the world. The parasite has a highly flexible life cycle, with three very different routes of transmission: ingestion of oocysts shed from the definitive host (wild and domestic cats), ingestion of bradyzoites through the consumption of infected tissues of intermediate hosts (any endothermic vertebrate) and vertical transmission of tachyzoites to the offspring of intermediate hosts. Infection with *T. gondii* in people and other intermediate hosts can lead to a wide spectrum of disease states, ranging from chronic, asymptomatic infection that may be associated with altered behaviour (Lagrue and Poulin, 2010) to severe, often fatal illness (Dubey and Jones, 2008).

The association between *T. gondii* and behaviour in general is an intriguing one. The idea that a parasite or pathogen can 'hijack' the mind and control the behaviour of its host is bound to elicit interest, particularly if the host in question is a human. Anecdotal evidence suggests that children infected with *T. gondii* have lower IQ and learning difficulties (Witting, 1979 and references therein), and a number of observational studies have suggested that people infected with *T. gondii* have specific sets of behavioural traits that differ from uninfected people (Flegr, 2007; Flegr et al., 2002, 2003; Havlíček et al., 2001). If this parasite does in fact cause these behavioural changes, and impair learning in children, this is an important public health concern and a reason to better educate the public about how to avoid *T. gondii* infection.

The significance of understanding behavioural changes induced by *T. gondii* is not limited to human health. It is widely acknowledged that behavioural changes in rodents infected with *T. gondii* are adaptive for the parasite because they appear to increase predation of infected rodents by cats, thus ensuring transmission of the parasite to its definitive host (Prandovszky et al., 2011; Webster, 2001, 2007; Webster and McConkey, 2010). One of the main arguments to support the idea that behavioural changes are adaptive manipulation by the parasite (rather than a simple by-product of infection) is that the behavioural changes observed are specific to those that would enhance transmission to cats (McConkey et al., 2013; Vyas, 2013; Webster et al., 2013). We have previously questioned the evidence for adaptive manipulation of intermediate hosts by *T. gondii* (Worth et al., 2013), partly because observed behavioural changes do not appear to be consistent across different studies.

While such inconsistencies have also been mentioned previously by other authors (Gonzalez et al., 2007; Skallová et al., 2006; Vyas, 2013; Webster et al., 2013), they have never previously been documented in any detail. Therefore, it is the explicit purpose of this review to provide a detailed summary and review of behavioural changes in laboratory mice and rats infected with *T. gondii*—with particular attention to contrasting results across studies. We will also discuss how inconsistencies in experimental design may explain some of these contrasting results, highlight some issues regarding the interpretation of behavioural studies and provide suggestions for how these issues may be addressed in future studies. Our aim is not to answer the question of whether behavioural changes seen in rodents infected with *T. gondii* are or are not an example of adaptive manipulation, but to indicate areas where we believe further work is required in order to provide a solid foundation for understanding the proximate and ultimate causes of these behavioural changes.

2. WHAT IS THE EVIDENCE FOR *T. GONDII*-INDUCED BEHAVIOURAL CHANGES?

2.1. Observational studies

Studies concerning the effect of *T. gondii* on human behaviour and mental health are by necessity observational studies, where personality and behavioural traits are compared between infected and uninfected people. There is evidence that infected humans have different personality profiles (Flegr et al., 2000, 2003), have longer reaction times (Havlíček et al., 2001) and are more likely to be involved in a traffic accident (Flegr et al., 2002) compared to uninfected humans (see Flegr, 2013 for a recent review). It has also been reported that there is an association between *T. gondii* infection and schizophrenia in humans (Torrey et al., 2007).

Observational studies of behavioural changes have also been conducted on other host species. Webster et al. (1994) trapped wild rats from farms, allocated them into groups based on serological diagnosis of the presence or absence of *T. gondii* infection and compared behavioural traits of infected and uninfected rats. Under certain conditions, they found an association between the presence of *T. gondii* infection and the presence of lower neophobia (measured by novel food consumption and tendency to enter traps) and proposed that this lowered neophobia may render infected rats more susceptible to predation by cats.

A fundamental problem with all observational studies is the difficulty in determining causation (see Taboas and McKay, 2012). If a correlation is found between the presence of naturally acquired *T. gondii* infection and certain behavioural traits in a host, does this mean that the parasite caused the change in behaviour or that the presence of these behavioural traits predisposed certain individuals to acquiring *T. gondii* infection? For example, Hutchison et al. (1980a) suggested that children who have learning difficulties may also be slow to acquire hygiene habits, making them more prone to acquiring *T. gondii* infection via contact with infective cat faeces.

For humans, some studies have attempted to overcome this limitation in determining causation by examining the relationship between personality changes and time since infection. Flegr et al. (2000) found a positive correlation between duration of infection and personality changes in women and suggested that this supports the idea that *T. gondii* infection causes the personality change, rather than the other way around. For laboratory rodents, researchers are able to use experimental infections to study in more detail the causal relationship between *T. gondii* infection and behaviour.

2.2. Experimental studies

Experimental studies may provide more robust evidence for behavioural changes induced by *T. gondii*. In rodents, these are usually conducted by experimentally infecting one group of rodents with *T. gondii* and 'sham infecting' a control group with saline. Particular behavioural traits are then measured and compared between groups. In the following sections, we summarise the results from experimental studies that have examined, in rodents, the effect of *T. gondii* infection on motor coordination, memory, learning, activity level, anxiety, response to novelty and response to cat odours.

2.2.1 Motor coordination

Motor coordination has been measured in mice infected with *T. gondii* using a number of different methods: the rotating cylinder test, balance beam test, gait analysis, righting reaction and rod test (Table 3.1).

Hutchison et al. (1980b) and Hay et al. (1983a) found that infected mice fell more often from a rotating cylinder than control mice, and Gulinello et al. (2010) found that infected mice had impaired motor coordination as indicated by an increased number of slips on the balance beam, an increased latency to cross the balance beam and abnormal gait characteristics. By contrast, Goodwin et al. (2012) found no effect of *T. gondii* infection on

Table 3.1 A summary of past studies of motor performance of *T. gondii*-infected mice

Effect on motor performance	Behaviour test	Host	*T. gondii* strain*	Weeks post infection	Reference
Impaired	Rotating cylinder	Mouse, male + female	Beverley[a]	26	Hutchison et al. (1980b)
Impaired	Rotating cylinder	Mouse, male + female	Beverley[a,c]	14	Hay et al. (1983a)
Impaired	Balance beam	Mouse, male	ME49[a]	7	Gulinello et al. (2010)
Impaired	Gait analysis	Mouse, male	ME49[a]	7	Gulinello et al. (2010)
None	Righting reaction	Mouse, male + female	VEG[c]	4, 8	Goodwin et al. (2012)
None	Rod test	Mouse, male + female	VEG[c]	4, 8	Goodwin et al. (2012)

*Superscript letters refer to type of *T. gondii* infection used; a = adult-acquired, c = congenital.

motor coordination in mice, as measured by the ability of a mouse to right itself after being placed on its back and how well a mouse could hold on to a rod.

Hay et al. (1983a) found that motor coordination was slightly more affected in congenitally infected mice than mice infected as adults; however, this was only statistically significant for infected mice that were the offspring of acutely infected mothers (as opposed to chronically infected mothers). Given that the motor performance of uninfected littermates from *T. gondii*-infected mothers was unaffected, it seems unlikely that maternal influences *per se* could explain any difference between the motor performance of congenitally infected mice and that of adult-infected mice. Hay et al. (1983a) suggested that the difference may be explained by suppressed immunity during pregnancy allowing more *T. gondii* tachyzoites to cross the placenta, leading to a larger number of parasites invading the developing foetus.

2.2.2 Learning and memory

Learning performance and memory have been measured in a number of studies of mice and rats infected with *T. gondii*, using a variety of techniques, mostly involving the ability to navigate a maze (Tables 3.2 and 3.3).

Witting (1979) used a 'deep maze' and learning was measured by how well animals navigated the maze over multiple trials; they found that the learning performance of both rats and mice was impaired in *T. gondii*-infected animals. Hodkova et al. (2007) found that infected mice performed worse in an eight-arm radial maze and a static rod learning test, although the authors suggested that this result may be better interpreted as evidence of a deficit in recognising novel stimuli, rather than true deficits in learning capacity. By contrast, Vyas et al. (2007a) found that *T. gondii* infection had no effect on the learning performance of rats tested in a Morris water maze, and Goodwin et al. (2012) found no learning impairment in mice in a Barnes maze test.

Witting (1979) tested memory by training uninfected animals in a deep maze and infecting them with *T. gondii* after they had learned the maze. Infected mice appeared to 'forget' how to navigate the maze, even though they had learned it prior to infection. Impaired memory for maze navigation was also found in mice by Kannan et al. (2010) and Goodwin et al. (2012), although Goodwin et al. (2012) found that memory impairment was specific to male mice at 8 weeks post infection; when tested 4 weeks post infection, male mice actually exhibited enhanced memory compared to

Table 3.2 A summary of past studies of the effect of *T. gondii* infection on learning performance in mice and rats

Effect on learning performance	Behaviour test	Host	*T. gondii* strain*	Weeks post infection	Reference
Impaired	Deep maze	Mouse, female	Weiss[a]	Multiple	Witting (1979)
Impaired	Deep maze	Rat, female	Weiss[a]	Multiple	Witting (1979)
Impaired	Static rod—time to cross	Mouse, female	HIF[a]	10	Hodkova et al. (2007)
Impaired	Eight-arm radial maze—# errors	Mouse, female	HIF[a]	10	Hodkova et al. (2007)
None	Morris water maze	Rat, male	PRU[a]	4	Vyas et al. (2007a)
None	Barnes maze test task acquisition	Mouse, male +female	VEG[c]	4, 8	Goodwin et al. (2012)

*Superscript letters refer to type of *T. gondii* infection used; a = adult-acquired, c = congenital.

uninfected mice. Xiao et al. (2012) found impaired olfactory memory in male mice, but not female mice.

In contrast to the studies showing memory impairment in mice, Kannan et al. (2010) found no effect of *T. gondii* infection (with the PRU strain) on spontaneous alternation behaviour or tendency to enter familiar versus novel arms of a Y maze. Spontaneous alternation behaviour refers to the tendency for mice and rats to alternate their choice of Y-maze arms on successive opportunities, and this is thought to be suitable for testing memory as an animal must remember which arm it previously entered in order to enter a 'new' arm (Hughes, 2004). Gulinello et al. (2010) found that infection had no effect on memory (spatial or recognition) or response to novel objects (object recognition and object placement tests).

Witting (1979), with the same deep-maze retention technique as used for mice, found no impairment of memory in infected rats, and Vyas et al. (2007a) also found that *T. gondii* infection had no effect on memory of rats, as measured in the Morris water maze.

Table 3.3 A summary of past studies of memory of *T. gondii*-infected mice and rats

Effect on memory	Behaviour test	Host	*T. gondii* strain*	Weeks post infection	Reference
Impaired	Deep-maze retention	Mouse, female	Weiss[a]	8	Witting (1979)
Impaired	Spontaneous alternations in Y maze	Mouse, female	ME49[a]	8	Kannan et al. (2010)
Impaired	Barnes maze short-term memory	Mouse, male	VEG[c]	8	Goodwin et al. (2012)
Impaired	STFP*—olfactory memory	Mouse, male	PRU[a]	8–16	Xiao et al. (2012)
Improved	Barnes maze short-term memory	Mouse, male	VEG[c]	4	Goodwin et al. (2012)
None	Deep-maze retention	Rat, female	Weiss[a]	8	Witting (1979)
None	Morris water maze	Rat, male	PRU[a]	4	Vyas et al. (2007a)
None	Novel object recognition	Mouse, male	ME49[a]	7	Gulinello et al. (2010)
None	Object placement	Mouse, male	ME49[a]	7	Gulinello et al. (2010)
None	Novel versus familiar arm of Y maze	Mouse, female	PRU[a]	8	Kannan et al. (2010)
None	Spontaneous alternations in Y maze	Mouse, female	PRU[a]	8	Kannan et al. (2010)
None	Novel versus familiar arm of Y maze	Mouse, female	ME49[a]	8	Kannan et al. (2010)
None	Barnes maze short-term memory	Mouse, female	VEG[c]	4, 8	Goodwin et al. (2012)

*Superscript letters refer to type of *T. gondii* infection used; a = adult-acquired, c = congenital. STFP = social transmission of food preference.

2.2.3 Locomotion

The effect of *T. gondii* on locomotion of mice and rats has been measured in terms of activity level and patterns of movement. Various methodologies have been employed, including open field-type arenas, different types of maze and exercise wheels. Studies investigating changes in activity level in *T. gondii*-infected mice and rats have reported contrasting results (Table 3.4). Some report that *T. gondii*-infected animals are less active (Gonzalez et al., 2007; Gulinello et al., 2010; Hrdá et al., 2000; Skallová et al., 2006; Witting, 1979), some report that they are more active (Afonso et al., 2012; Hay et al., 1983b, 1984; Hodkova et al., 2007; Kannan et al., 2010; Webster, 1994) and the remainder report no difference between infected and uninfected rodents (Gonzalez et al., 2007; Hrdá et al., 2000; Kannan et al., 2010; Vyas et al., 2007a).

As well as 'activity level', *T. gondii* appears to have some effect on patterns of movement of infected rodents, although contrasting results exist. Hutchison et al. (1980c) reported that infected mice exhibit an increased number of shorter bouts in the open field, while Afonso et al. (2012) found the opposite, with infected mice showing a decrease in the frequency of shorter bouts and an increase in the frequency of long and very long bouts. Vyas et al. (2007a) examined the locomotion of rats in the open field and found no difference in the number of progression segments of infected versus uninfected rats.

2.2.4 Anxiety

Anxiety has been measured using the elevated plus maze (EPM), the social interaction test (SIT) and the open field. Decreased anxiety is indicated by increased time spent on open arms in the EPM (Walf and Frye, 2007), increased time spent in social interaction in the SIT (File and Hyde, 1978) and increased time spent in the centre of the open field (Walsh and Cummins, 1976). As also mentioned by Webster and McConkey (2010), there is inconsistency in the reported effect of infection with *T. gondii* on anxiety behaviour in mice and rats (Table 3.5).

With infected mice, most studies have interpreted changes in anxiety based on behaviour in the open field. Hay et al. (1983b, 1984) and Gatkowska et al. (2012) found that infected mice spent less time than control mice in the central squares of the open field, indicating an increase in anxiety. Skallová et al. (2006) reported the same result for infected female mice but found that infected male mice exhibited a decrease in anxiety in the open field. Afonso et al. (2012) found no change in open-field anxiety in infected mice, but reported a decrease in anxiety when mice were tested in the EPM.

Gonzalez et al. (2007) found a decrease in anxiety in rats infected with 100 or 1000 *T. gondii* tachyzoites, using EPM and SIT methodologies; this decrease in anxiety was evident at 3 and 7 weeks post infection in the EPM,

Table 3.4 Summary of past studies that have measured activity level in *T. gondii*-infected mice and rats

Effect on activity level	Behaviour test	Host	*T. gondii* strain* (dose†)	Weeks post infection	Reference
Decreased	Deep maze	Mouse, female	Weiss[a]	Multiple	Witting (1979)
Decreased	Deep maze	Rat, female	Weiss[a]	Multiple	Witting (1979)
Decreased	Open-field squares entered	Mouse, male + female	01529/38[a]	3	Hrdá et al. (2000)
Decreased	Open-field squares entered	Mouse, female	HIF[a]	10	Skallová et al. (2006)
Decreased	Social interaction test	Rat, unknown sex	RH[a] (1500)	7	Gonzalez et al. (2007)
Decreased	Open-field grid crosses	Mouse, male	ME49[a]	7	Gulinello et al. (2010)
Decreased	Open-field beams broken	Mouse, male	PRU[a]	8–16	Xiao et al. (2012)
Increased	Open-field squares entered	Mouse, male + female	Beverley[a,c]	14	Hay et al. (1983b)
Increased	Open-field squares entered	Mouse, male + female	Beverley[c]	14	Hay et al. (1984)
Increased	1–0 Sampling 'active outside nest box'	Rat, male + female	Beverley[a]	Unknown	Webster (1994b)
Increased	1–0 Sampling 'active outside nest box'	Rat, female	Beverley[c]	Unknown	Webster (1994b)
Increased	Spontaneous wheel running distance and velocity	Mouse, female	HIF[a]	12	Hodkova et al. (2007)

Table 3.4 Summary of past studies that have measured activity level in *T. gondii*-infected mice and rats—cont'd

Effect on activity level	Behaviour test	Host	*T. gondii* strain (dose)	Weeks post infection	Reference
Increased	Open-field beams broken	Mouse, female	PRU[a]	8	Kannan et al. (2010)
Increased	Open-field distance travelled and average speed	Mouse, female	ME49[a]	9	Afonso et al. (2012)
Increased	EPM distance travelled	Mouse, female	ME49[a]	9	Afonso et al. (2012)
Increased	Open-field beams broken	Mouse, female	PRU[a]	8–16	Xiao et al. (2012)
None	1–0 Sampling 'active outside nest box'	Rat, male	Beverley[c]	Unknown	Webster (1994b)
None	Open-field squares entered	Mouse, male + female	01529/38[a]	6, 12	Hrdá et al. (2000)
None	Open-field squares entered	Mouse, male	HIF[a]	10	Skallová et al. (2006)
None	EPM closed-arm entries	Rat, unknown sex	RH[a] (100, 1000, 1500)	3, 7	Gonzalez et al. (2007)
None	Social interaction test	Rat, unknown sex	RH[a] (100, 1000, 1500)	3	Gonzalez et al. (2007)
None	Social interaction test	Rat, unknown sex	RH[a] (100, 1000)	7	Gonzalez et al. (2007)
None	Spontaneous wheel running time spent running	Mouse, female	HIF[a]	12	Hodkova et al. (2007)

Continued

Table 3.4 Summary of past studies that have measured activity level in *T. gondii*-infected mice and rats—cont'd

Effect on activity level	Behaviour test	Host	*T. gondii* strain (dose)	Weeks post infection	Reference
None	Open-field distance travelled and maximal speed	Rat, male	PRU[a]	4	Vyas et al. (2007a)
None	Open-field beams broken	Mouse, female	ME49[a]	8	Kannan et al. (2010)
None	Barnes maze # holes visited	Mouse, male + female	VEG[c]	4, 8	Goodwin et al. (2012)

*Superscript letters refer to type of *T. gondii* infection used; a = adult-acquired, c = congenital.
†Dose is provided in parentheses when relevant to interpretation of table.

but only at 7 weeks post infection in the SIT. Rats infected with a higher dose of parasites (1500 tachyzoites) did not show a decrease in anxiety in either test, possibly because a change in anxiety was masked by a decrease in activity level in this group. In contrast to Gonzalez et al. (2007), Vyas et al. (2007a) found no change in anxiety level of rats as measured on the EPM or in the open field. Despite using a very high dose of parasites, Vyas et al. (2007a) did not detect any change in activity level in the open field, so it is unlikely that a decrease in anxiety was masked by changes in activity in this case.

2.2.5 Response to novelty

The response of rodents to novelty has been tested using novel environments (Y maze and open field) and introducing novel objects into a familiar environment. Only a small number of studies have been conducted and these have not found a consistent response (Table 3.6).

Hutchison et al. (1980a) and Hay et al. (1983b) found decreased exploration of novel areas in infected mice in maze and open-field tests. Hutchison et al. (1980c) investigated the response of mice to a novel object placed in their environment and found that, whereas uninfected mice reacted strongly to the object initially, and habituated over time, infected mice did not respond as markedly when the object was first introduced and did not appear to habituate. Gulinello et al. (2010), however, reported

Table 3.5 Summary of past studies of anxiety level in *T. gondii*-infected mice and rats

Effect on anxiety	Behaviour test	Host	*T. gondii* strain* (dose†)	Weeks post infection	Reference
Decreased anxiety	Open-field central square entries	Mouse, male	HIF[a]	10	Skallová et al. (2006)
Decreased anxiety	SIT time spent in social investigation	Rat, unknown	RH[a] (100, 1000)	7	Gonzalez et al. (2007)
Decreased anxiety	EPM open-arm time/entries	Rat, unknown	RH[a] (100, 1000)	3, 7	Gonzalez et al. (2007)
Decreased Anxiety	EPM open-arm end visits	Mouse, female	ME49[a]	9	Afonso et al. (2012)
Increased anxiety	Open-field time spent centre	Mouse, male + female	Beverley[a,c]	14	Hay et al. (1983b)
Increased anxiety	Open-field time spent centre	Mouse, male + female	Beverley[c]	14	Hay et al. (1984)
Increased anxiety	Open-field central square entries	Mouse, female	HIF[a]	10	Skallová et al. (2006)
Increased anxiety	Open-field time spent centre	Mouse, male	ME49[a]	3, 6	Gatkowska et al. (2012)
No difference	SIT time spent in social investigation	Rat, unknown	RH[a] (100, 1000)	3	Gonzalez et al. (2007)
No difference	EPM open-arm time/entries	Rat, unknown	RH[a] (1500)	7	Gonzalez et al. (2007)
No difference	SIT time spent in social investigation	Rat, unknown	RH[a] (1500)	7	Gonzalez et al. (2007)
No difference	Open-field time spent centre	Rat, male	PRU[a]	4	Vyas et al. (2007a)
No difference	Open-field time spent centre	Mouse, female	ME49[a]	9	Afonso et al. (2012)

*Superscript letters refer to type of *T. gondii* infection used; a = adult-acquired, c = congenital.
†Dose is provided in parentheses when relevant to interpretation of table.

Table 3.6 Summary of past studies concerning response to novelty by *T. gondii*-infected mice and rats

Exploration of novelty	Behaviour test	Host	*T. gondii* strain*	Weeks post infection	Reference
Decreased	Y-maze time in novel versus familiar arm	Mouse, male	Beverly[a]	8	Hutchison et al. (1980a)
Decreased	Preference for central area of open field	Mouse, male + female	Beverly[c,a]	14–15	Hay et al. (1983b)
Abnormal	Response to novel object	Mouse, unknown	Beverly[a]	8	Hutchison et al. (1980c)
Normal	Response to novel objects	Mouse, male	ME49[a]	7	Gulinello et al. (2010)
Increased	# Approaches to human observer	Rat, male + female	Beverly[c]	22–29	Berdoy et al. (1995)

*Superscript letters refer to type of *T. gondii* infection used; a = adult-acquired, c = congenital.

that infected mice spent the same amount of time as uninfected mice exploring novel objects in their environment.

Berdoy et al. (1995) found that infected rats were more likely than uninfected rats to approach a human observer.

2.2.6 Aversion to feline odour

One of the most publicised effects of *T. gondii* on rodent behaviour is that it causes rodents that are normally fearful of cats to become attracted to cat odour. Cat odour avoidance has been measured in different arenas (circular, rectangular and square of different sizes) by presenting various odours in different areas of the arena and recording time spent near each odour (Table 3.7).

Berdoy et al. (2000) were the first to report this attraction to cat odour; they found that infected rats spent more time than uninfected rats near cat urine, whereas infected and uninfected rats spent comparable amounts of time near other odours (their own, rabbit urine and water) indicating that the differential response of uninfected versus infected rats was specific to cat urine.

Table 3.7 Summary of past studies of response of *T. gondii*-infected mice and rats to feline odours

Effect on aversion to cat odour	Behaviour test	Host	*T. gondii* strain*	Weeks post infection	Reference
Aversion decreased	Cat urine, square pen with brick maze	Rat, unknown	Beverley[a]	Unknown	Berdoy et al. (2000)
Aversion decreased	Bobcat urine, rectangular arena	Mouse, female	PRU[a]	4	Vyas et al. (2007a)
Aversion decreased	Cat fur, rectangular arena	Mouse, female	PRU[a]	4	Vyas et al. (2007a)
Aversion decreased	Bobcat versus rabbit urine, circular arena	Rat, male	PRU[a]	4	Vyas et al. (2007a)
Aversion decreased	Bobcat urine, rectangular arena	Rat, male	PRU[a]	7–8	Vyas et al. (2007b)
Aversion decreased	Cat fur, rectangular arena	Rat, male	PRU[a]	7–8	Vyas et al. (2007b)
Aversion decreased	Cat versus mink urine, Y maze	Rat, unknown	ME49[a]	Unknown	Lamberton et al. (2008)
Aversion decreased	Cat urine, square arena	Mouse, female	ME49[a]	8	Kannan et al. (2010)
Aversion decreased	Cat urine, square arena	Mouse, female	PRU[a]	8	Kannan et al. (2010)
Aversion decreased	Bobcat versus mink urine, three-chambered arena	Mouse, female	PRU[a]	8–16	Xiao et al. (2012)
No change in aversion	Cat fur versus unscented towel, circular arena	Rat, male	PRU[a]	4	Vyas et al. (2007a)
No change in aversion	Cat fur, rectangular arena	Rat, male	PRU[a]	7–8	Vyas et al. (2007b)
No change in aversion	Cat urine, square arena	Mouse, female	ME49[a]	28	Kannan et al. (2010)
No change in aversion†	Cat urine, square arena	Mouse, female	PRU[a]	28	Kannan et al. (2010)
No change in aversion	Bobcat versus mink urine, three-chambered arena	Mouse, male	PRU[a]	8–16	Xiao et al. (2012)

*Superscript letters refer to type of *T. gondii* infection used; a = adult-acquired, c = congenital.
†The authors reported that PRU-infected mice are still attracted to cat urine at 7 months post infection. However, the difference between PRU-infected and controls is not significant (p value is 0.066).

Vyas et al. (2007a) tested the response of uninfected and infected rats to bobcat urine *relative* to rabbit urine. Uninfected rats spent 10.4% of their time near bobcat urine and 23% near rabbit urine, while infected rats spent slightly more time near bobcat urine (16.6%) and less time near rabbit urine (9.9%). These authors did not directly compare time spent near bobcat urine for uninfected versus infected rats, but compared the 'occupancy ratio' (time spent near bobcat urine/(time spent near bobcat urine + time spent near rabbit urine)) for uninfected and infected rats. They found that the occupancy ratio was much higher for infected rats and suggested that this was the evidence of attraction to bobcat urine. However, there are possible limitations to this interpretation given that the occupancy ratio is dependent on 'time spent near rabbit urine', which was very different for uninfected and infected rats.

Vyas et al. (2007a) studied the response of female mice to bobcat odour in the same circular arena as they used for rats. The results for mice are more convincing than for rats, although again the time spent near bobcat urine was not directly compared between infected and uninfected groups. Uninfected mice spent 55% time near rabbit urine and only 5% of time near bobcat urine, while infected mice spent 35% of time near rabbit urine and 25% of time near bobcat urine. Xiao et al. (2012) conducted a similar investigation of the relative response of uninfected and infected male and female mice to bobcat urine compared to mink urine. These authors report that aversion to feline odour was decreased only in female mice and not in male mice.

In agreement with Vyas et al. (2007a), Kannan et al. (2010) reported convincing evidence of mice losing their fear of domestic cat urine. Infected mice spent approximately twice as much time as uninfected mice near cat urine, yet did not differ in the amount of time spent near dog urine. However, this altered response to cat odour was time-dependent, being evident 2 months after infection, but not 7 months after infection.

Aversion to a different sort of cat odour (fur) was tested by Vyas et al. (2007a) in both mice and rats, with a differential response between species. Whereas both infected and uninfected rats showed strong avoidance of cat fur odour, mice differed in their response to cat fur depending on infection with *T. gondii*. Uninfected mice spent 40% of time in the cat fur bisect and infected mice spent 75% of their time in the cat fur bisect, suggesting that in this case, the ambivalence (rather than avoidance) of uninfected mice to cat fur should be contrasted with an attraction in infected mice.

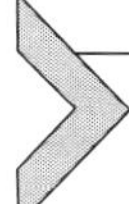

3. POSSIBLE REASONS FOR INCONSISTENCIES IN PAST STUDIES

Past studies of behavioural changes induced by *T. gondii* have used a range of different host species, including different strains of inbred and outbred mice and different strains of laboratory and wild (or hybrid) rats. Multiple strains of *T. gondii* have been used, including HIF, RH, BEV, PRU and ME49, and different infective stages (tachyzoites and bradyzoites) have been inoculated into experimental subjects using either oral gavage, subcutaneous injection or intraperitoneal injection. To add to the variety of experimental factors, past studies have used different methodologies to test behaviour and have measured behaviour at different times after infection, ranging from 3 to 28 weeks post infection, with the majority testing behaviour at 8–10 weeks post infection. All of these experimental variables may influence the behavioural outcome of infection with *T. gondii* and may therefore account for the inconsistent results seen among different studies. Webster et al. (2013) also provided a detailed discussion of how experimental factors may influence the outcome (behaviourally and clinically) of *T. gondii* infection and, in doing so, made recommendations concerning experimental designs (e.g. host species, *T. gondii* strain and type of behaviour test) for future behavioural studies. In the following discussion, we consider which factors are most likely to account for the contrasting results observed in the behavioural studies detailed in Tables 3.1–3.6.

The least ambiguous way to determine whether an experimental factor influences results of a behavioural test would be to manipulate it while controlling for all other variables. While most past studies have only investigated *T. gondii*-induced behavioural changes under one set of conditions, a number have directly compared the effects of different *T. gondii* strains, different host sexes and species and acute versus chronic infections and therefore do provide some insight into possible reasons for inconsistent results.

3.1. Differences between *T. gondii* strain and dose

T. gondii strain and/or dose-dependent differences in host immune activation and parasite growth and dissemination have been reported in animal models (Araujo et al., 1976; Darde et al., 2007; Saeij et al., 2005). There are also reported differences in parasite dissemination depending on the route of administration used to infect experimental animals (Boyle et al., 2007). These 'parasite inoculum'-dependent differences in disease outcome and dissemination of the parasite may extend to different effects on behaviour.

Kannan et al. (2010) found that the effect of infection with *T. gondii* on activity and memory of mice was strain-dependent. Mice infected with the PRU strain of *T. gondii* showed increased activity level and normal working memory (spontaneous alternations in Y maze), while mice infected with ME49 showed normal levels of activity but impaired working memory (although the *p* value for comparison of spontaneous alterations by ME49-infected mice and control mice did not quite reach significance ($p=0.056$), which may be a consequence of the small sample size (control $n=10$ and ME49-infected $n=8$)).

Gonzalez et al. (2007) found evidence for different behavioural changes depending on the initial dose of tachyzoites inoculated into rats. Rats infected with 100 or 1000 tachyzoites showed a decrease in anxiety but no change in overall activity level (compared to uninfected rats) when tested in EPM and SITs. In the EPM, the decrease in anxiety was dose-dependent; rats infected with 1000 tachyzoites showed a bigger decrease in anxiety than those infected with 100 tachyzoites. In contrast, rats infected with 1500 tachyzoites showed a drop in locomotion and no change in anxiety level. These differences possibly reflect lethargy due to sickness in the rats inoculated with the highest dose.

3.2. Differences between host species

Different hosts have different susceptibility to *T. gondii* infection, with rats generally being more resistant or tolerant to toxoplasmosis than mice (Innes, 1997). There is also evidence of differences in susceptibility between strains of mice (Araujo et al., 1976) and rats (Kempf et al., 1999). Webster et al. (2013) suggested that rats may be better models (than mice) for human toxoplasmosis as the outcome of infection is more similar to that in humans. These differences in susceptibility to disease may extend to differences in behavioural changes, particularly if behavioural changes are related to sickness and/or number of cysts formed in the brain. Alternatively, differences in *T. gondii*-induced behavioural changes between mice and rats may exist due to different baseline behaviours of each species.

Few studies have directly compared the behavioural manifestation of *T. gondii* infection between mice and rats, but studies that have compared these hosts found differences between them. Witting (1979) reported that while learning was impaired in both host species, memory was impaired only in mice and not in rats. Vyas et al. (2007a) studied the response of both mice and rats to cat fur and urine odours. Infected individuals of both species showed an increase in time spent near bobcat urine, but only infected mice showed the same increase in response to cat fur odour. It is not possible to determine whether differences between mice and rats were due to different levels of immune activation and/or number of cysts in the brain, as neither study measured these variables.

Although response to novelty has not been directly compared in mice and rats, it has been suggested, based on qualitative differences (Table 3.6), that *T. gondii* infection affects this behaviour differently in the two species (Gonzalez et al., 2007; Hodkova et al., 2007). Infected mice appear to become less interested in novelty, while infected rats become more interested in novelty (Table 3.6). This may represent a reversal of normal behaviour for each species; as mentioned by Hodkova et al. (2007), mice are usually considered neophilic, while rats are considered neophobic. If so, this would argue against a mechanism that causes a general reduction in fear or anxiety. Further research, including direct comparisons between mice and rats, will improve our understanding of differences between host species and is likely to provide clues about the mechanism by which *T. gondii* induces behavioural changes.

3.3. Difference between male and female hosts

Males and females might be expected to be affected differently by *T. gondii* infection because they differ in susceptibility to disease caused by the parasite (Roberts et al., 1995) and also differ naturally in certain behaviours. For example, for uninfected control mice in past studies, females tend to be more active in the open field than males (Hay et al., 1983b) and males tend to fall more often than females from the rotating cylinder (Hay et al., 1983a; Hutchison et al., 1980b). Uninfected male rats are more inquisitive than uninfected female rats (Berdoy et al., 1995).

Despite these inherent differences between male and female rodents, many studies have found that the parasite has the same effect on each sex, with regard to activity level (Goodwin et al., 2012; Hay et al., 1983b, 1984; Hrdá et al., 2000), motor coordination (Goodwin et al., 2012; Hay et al., 1983a; Hutchison et al., 1980b), learning performance (Goodwin et al., 2012) and time spent in central part of the open field (Hay et al., 1983b, 1984).

A number of studies, however, have found sex-dependent differences in behavioural changes induced by *T. gondii*. Skallová et al. (2006) reported a decrease in activity level for infected female mice, while activity level of infected male mice remained normal. Goodwin et al. (2012) found that memory was not affected by infection in female mice, but was affected in male mice.

Xiao et al. (2012) studied changes in gene expression as well as changes in behaviour in male and female mice infected with *T. gondii*. They found that infected female mice exhibited changes in the expression of genes related to forebrain development, neurogenesis and sensory and motor coordination, while infected male mice exhibited changes in the expression of genes related to olfactory function. With regard to behavioural changes, Xiao et al. (2012) found that only infected female mice showed an altered response to cat urine, as well as increased activity in the open-field and intact olfactory memory, while infected male mice showed a normal response to cat urine, decreased activity in the open-field and impaired olfactory memory.

3.4. Differences between measurement times

The effect of *T. gondii* on behaviour may depend on the time post infection, particularly if behavioural changes are related to replication of the parasite, or host immune response, both of which will differ depending on whether the animal is experiencing acute or chronic infection. Several examples from past studies support the idea that the type of behavioural changes observed may depend on time post infection. For example, Hrdá et al. (2000) found a decrease in activity at 3 weeks post infection, which coincided with maximal symptoms of sickness during acute infection. At 6 and 12 weeks post infection, infected mice no longer differed from control mice, suggesting that the decrease in activity at 3 weeks post infection was transient and most likely due to sickness of the animals. Goodwin et al. (2012) found that most of the behaviours they studied were not affected by *T. gondii* infection, and this was consistent across time points (4 and 8 weeks post infection). The exception was memory of infected male mice, which deviated from controls in opposite directions at 4 weeks compared to 8 weeks post infection (Table 3.3). Gonzalez et al. (2007) found that a decrease in anxiety in the SIT was only detectable at 7 weeks post infection. However, at 3 weeks post infection, a decrease in anxiety was measurable on the EPM, so the lack of effect in the SIT may simply be a consequence of different behaviours being measured by different tests (see text later). Gatkowska et al. (2012) compared behavioural changes in the same group of mice during acute (3 weeks post infection) and chronic (6 weeks post infection) *T. gondii* infection and found some differences in behaviour at the two time points. However, differences at the two time points are difficult to interpret, as there was no control group to account for repeated exposure of mice to the behaviour testing apparatus.

Understanding when certain behavioural changes occur (and whether they are transient or permanent) will provide important clues to the mechanism(s) by which *T. gondii* affects behaviour. For example, if changes are transient and only occur during early chronic or acute infection, this suggests that they are likely to be related to host immune response and/or damage caused by the replication of the parasite. On the other hand, behavioural changes that occur long after the initial infection may argue for a more permanent mechanism related to the presence of the *T. gondii* tissue cysts in the host's body. It is likely that different behavioural changes are due to different mechanisms, and some may be transient while others are permanent. Therefore, in order to better understand this aspect of *T. gondii*-induced behavioural changes, it would be valuable for future studies to assess multiple types of behavioural changes at different time points post infection.

3.5. Measurement error

There are two main types of measurement error in biological studies, both of which may have contributed to inconsistencies in past studies of behavioural manipulation by *T. gondii*. The first type, known as reliability (Bollen, 1989), is what we usually think of as a mistake in the measurement of a response variable. For example, was the time spent by a rat in an open field accurately recorded? Were some slips on a balance beam missed because the animal recovered too rapidly? This type of error can often be minimised by rigorous quality control and large sample sizes but may represent a non-trivial problem in behavioural manipulation studies, because response variables such as righting ability, spontaneous alternation and occupancy ratio can be difficult to measure.

The second type of error, known as validity, refers to the degree to which a response variable matches the construct it seeks to measure (Carter et al., 2013). This leads to subjectivity in interpretation, since the responses recorded in behavioural tests may not measure what researchers think they measure. For example, although a decrease in spontaneous alternation in a Y maze is commonly interpreted as an impairment of memory retention, it may also reflect changes in attention, sensory acuity, encoding or motivation (Hughes, 2004). Similarly, it is possible that the rotating cylinder measures things other than motor performance/coordination. Impaired learning ability, decreased stamina or decreased muscular strength may also lead to an increase in the number of falls made by an infected animal. The nematode *Toxocara canis* causes a decrease in muscle weakness in rats (Chieffi et al., 2009), which is thought to be due to muscle tissue damage caused by the migration of the nematode larvae. To our knowledge, changes in muscular strength *per se* have not been measured for *T. gondii*-infected rodents. However, given that this parasite also invades muscle tissue (Remington and Cavanaugh, 1965), it may reduce muscular strength.

The validity of a behavioural test is also affected when the same label is given to behaviours that are measured using different methodologies. This is not a problem if the different methodologies actually do measure the same aspect of behaviour, but if they do not, it makes comparison between studies difficult. For example, activity level has been measured in a variety of ways in past studies of *T. gondii*-induced behavioural changes (Table 3.4), and it is likely that these different methodologies were measuring different aspects of behaviour. Webster (1994), for example, reported that infected rats were more active than uninfected rats, meaning that they were more likely to be out of the nest and moving in a complex environment. This may reflect a change in motivation to explore rather than a change in activity level *per se*. Other studies (Table 3.4) have measured distance travelled in an open field with no refuge, which is said to reflect novelty-induced activity. Some studies have used exercise wheels to measure activity level (Hodkova et al., 2007), which may detect differences in motivation to exercise and therefore represent a different behaviour to both distance travelled in the open field and motivation to explore.

Additionally, changes in one behaviour may confound the results of other behavioural tests. For example, behaviour in the EPM can be confounded if activity level is affected; a lethargic rodent may not move about enough to allow detection of a change in anxiety, even if one has occurred (e.g. Gonzalez et al., 2007).

While problems with the reliability of behavioural measurements can often be overcome through quality control during the experimental phase, issues of validity often require the development of multiple tests of behaviour, followed by multivariate data analysis to identify behavioural constructs (Carter et al., 2013). Structural equation modelling, a technique that allows the explicit incorporation of measurement errors and latent (unmeasured or theoretical) constructs, may be a particularly useful approach in behavioural manipulation studies, although it does require large sample sizes (Tomer and Pugesek, 2003).

3.6. What does this all mean for our understanding of *T. gondii*-induced behavioural changes?

When the results of past experimental studies are taken collectively, it becomes apparent that there is not a consistent, predictable effect of *T. gondii* on rodent behaviour. There are many inconsistencies in the results of behaviour tests from past studies, and *T. gondii* strain, dose, host species, host sex, time post infection and measurement apparatus can all influence the results of behaviour tests (Webster et al., 2013; Worth et al., 2013). This strongly suggests that while conclusions can be made for a particular study, the ability to extrapolate from one set of conditions to another is minimal. Further research is required to improve our understanding of how behavioural changes differ in response to different experimental conditions. This understanding will, in turn, generate useful hypotheses to guide research concerning the mechanism(s) by which *T. gondii*-induced behavioural changes occur.

4. INTERPRETATION OF THE EVIDENCE: ARE BEHAVIOURAL CHANGES ADAPTIVE FOR THE PARASITE?

Any changes in host behaviour that are induced by *T. gondii* may be adaptive for the parasite or may simply be a by-product of parasite infection. Parasite-induced changes in host behaviour can be considered to be adaptive if they increase parasite fitness through increased transmission (Poulin, 2010). Therefore, to properly interpret the significance of behavioural changes induced in rodents by *T. gondii*, we need to know, firstly, whether these changes increase the chances of predation by cats (or other hosts) and, secondly, whether increased predation by cats increases the chance of successful transmission of the parasite.

4.1. Do observed behavioural changes actually translate into increased predation by cats?

The behavioural changes observed in rodents infected with *T. gondii* are often interpreted as increasing the risk of predation and therefore enhancing transmission to cats, the definitive host for the parasite. For example, increased activity level may increase predation risk as cats are attracted by movement (Hay et al., 1983b). An abnormal response to (or inability to recognise) novel objects could also be argued to enhance predation risk as the affected rodents may be less responsive to the risk of predation (Hutchison et al., 1980a; Webster et al., 1994). Importantly, however, logical arguments can also be made for why these behavioural changes may not enhance predation by cats (Worth et al., 2013).

What is missing is the final step in this causal chain: experimental evidence that behavioural changes induced by *T. gondii* do increase predation rate of infected hosts by cats. While this evidence is difficult to obtain due to ethical reasons (Berdoy et al., 2000; Vyas and Sapolsky, 2010; Vyas et al., 2007a), the quantification of enhanced predation rate is a key requirement for accepting the adaptive nature of host manipulation by trophically transmitted parasites (Cézilly et al., 2010; Poulin, 2010).

4.2. Does cat predation increase transmission rate and parasite fitness?

The 'behavioural manipulation hypothesis' requires that the increased predation rate of infected intermediate hosts increases transmission success of the parasite (Cézilly et al., 2010). Simulation studies have shown that this may occur even with quite small increases in the susceptibility to predation of infected hosts (e.g. Vervaeke et al., 2006) and even when manipulation increases predation by non-target final hosts, as long as the initial predation risk for hosts is low (Seppälä and Jokela, 2008). The conclusions of these theoretical studies, however, cannot be easily generalised to *T. gondii*, which has a very complex and flexible life cycle.

T. gondii (unlike other Apicomplexa) can be transmitted from one host to another in a variety of ways, and predation of an infected rodent by a feline definitive host is not required for successful transmission. First, in addition to rodents, almost any other endothermic vertebrate can act as an intermediate host. Second, transmission can occur without involving the definitive host, through predation or scavenging of infected intermediate hosts by other intermediate hosts (Saeij et al., 2005), vertical transmission from infected mother to offspring (Johnson, 1997) and possibly sexual transmission from male to female (Arantes et al., 2009; Dass et al., 2012). The relative importance of different intermediate hosts and different transmission routes to the overall transmission success and therefore fitness of *T. gondii* is unknown.

Lélu et al. (2013) modelled the evolution of host manipulation by *T. gondii* and showed that selection for manipulation was affected by a number of factors, including epidemiological dynamics, virulence of the parasite and the existence of successful vertical transmission. This model represents a very useful start to studying the problem of whether host manipulation by *T. gondii* is adaptive, but it is a necessarily simplified representation of the complexity of transmission of this parasite. A more complex model of the role that enhanced predation by cats through behavioural manipulation of rodents plays in the successful transmission of the parasite would need to consider the interactions between cat predation on rodents and all other potential transmission routes (Fig. 3.1). This would require estimates, for example, of the

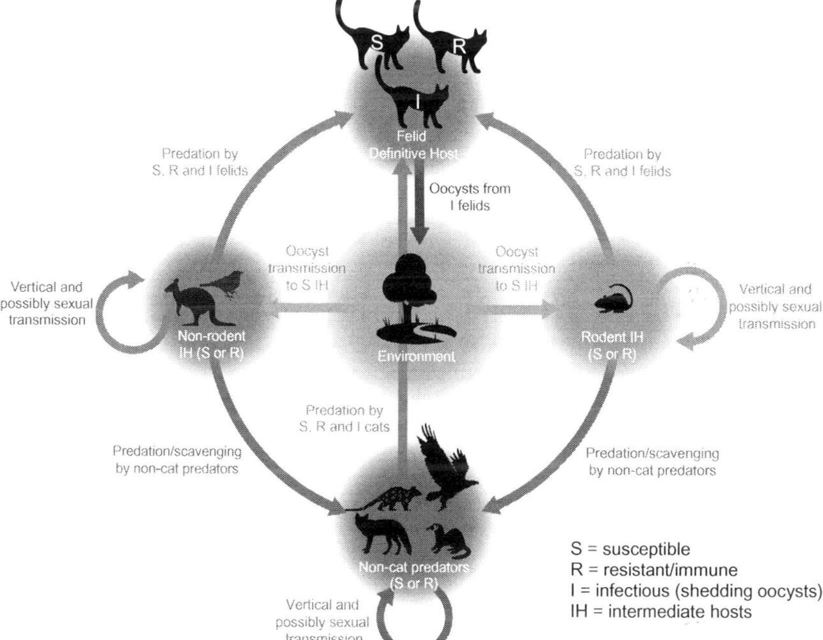

Figure 3.1 This figure aims to illustrate the complicated interrelationships between parameters that need to be considered in order to better understand the possible outcome of behavioural manipulation of rodent IH on overall transmission success and therefore fitness of *T. gondii*. Some examples to illustrate the complexity of the system are mentioned in the succeeding text: (1) Behavioural manipulation of rodent IH to increase transmission to cats may cause changes in other transmission pathways (e.g. an increase in predation by non-cat predators and/or a decrease in transmission between IHs by vertical or sexual transmission), which may affect the overall fitness of *T. gondii* in either a positive or a negative way, depending on the relative importance of each affected transmission pathway. (2) Behavioural manipulation 'designed' for rodent IH may also affect other IH species, and this may have a positive or negative impact on the overall success of *T. gondii* transmission, depending on the behavioural outcome of host manipulation in these other species. (3) Other as yet untested factors may be at play that may plausibly enhance overall transmission success of *T. gondii*—for example, if defecation behaviour of shedding, infectious cats is different to non-infected or resistant cats, this may benefit the parasite if oocysts are shed in a more favourable position (e.g. cool, damp areas).

relative infection rates of rodents and other intermediate hosts; the behavioural changes induced in this range of hosts; how these behavioural changes affect predation rate by cats and other predators; and how they may affect transmission through other routes such as scavenging, sexual contact and mother to offspring. In addition, it would be necessary to consider the effect of parasite strain and host species on parameters such as behavioural manipulation, virulence and success of vertical and sexual transmission.

5. PROXIMATE MECHANISMS OF BEHAVIOURAL CHANGE: WHAT DO PAST STUDIES SUGGEST?

Understanding the proximate mechanism underlying changes in host behaviour induced by infection with *T. gondii* is important as it will improve our understanding of how this parasite may impact human mental health, lead to treatments for behavioural/personality changes caused by the parasite and provide more evidence regarding the ultimate mechanisms (the evolutionary significance of behavioural changes).

Currently, the proximate mechanism(s) by which *T. gondii* induces behavioural changes in the host is not known, but there are three main (not mutually exclusive) possibilities (for review, see Kaushik et al., 2012):

1. Localisation of cysts in the brain (causing direct changes and/or damage to the brain)
2. Modulation of neurotransmitters and/or hormones that then alter behaviour
3. Modulation of immune response that then alters behaviour

In the following section, we summarise the experimental evidence from past studies for the first of these potential mechanisms. A discussion of the direct experimental evidence linking *T. gondii* behavioural changes with neurotransmitters, hormones and/or immune factors is limited, as very few studies have measured these molecules *and* also demonstrated behavioural changes. Although various lines of evidence support the hypothesis that the neurotransmitter dopamine may be involved in behavioural changes induced by *T. gondii* (see Kaushik et al., 2012), only one study that we are aware of has investigated behavioural changes and measured neurotransmitter levels within the same experiment (Goodwin et al., 2012). Given the inconsistencies in behavioural changes reported in past studies, future studies of neurotransmitter levels in *T. gondii*-infected rodents should also demonstrate changes in behaviour in their model of infection, so that changes in brain chemistry can be put into context.

5.1. Correlation between brain cyst number/location and severity of behavioural change?

There is some evidence to suggest that a larger number of cysts in the brains of infected rodents leads to a greater deviation from normal mouse behaviour. For example, Witting (1979) found that the total number of brain cysts correlated positively with the level of learning impairment in mice, while Hutchison et al. (1980a) found that brain cyst loads correlated negatively with time spent exploring the novel arm of a Y maze. In contrast, many studies have found no correlation between measures of behaviour and number of brain cysts (Hay et al., 1983a,b, 1984; Hutchison et al., 1980b), although in some cases, this may be due to the low variability in number of brain cysts in infected animals (Hutchison et al., 1980b). Afonso et al. (2012) found that some mice that were injected with *T. gondii* tachyzoites did not develop brain cysts. Some behaviours, such as increased activity level and an apparent lack of habituation to the open field, were exhibited only by animals that had developed a chronic *T. gondii* infection and therefore may be related specifically to the presence of cysts in the brain. However, all *T. gondii*-exposed mice, regardless of whether or not they had brain cysts, exhibited a decrease in time spent motionless in the open field. This suggests that the presence of brain cysts is not the only factor at play and different physiological mechanisms may underlie different types of behavioural changes. In short, the mechanisms responsible for *T. gondii*-induced behavioural changes are likely to be complex, and it does not appear that behavioural changes can be attributed solely to the number of cysts in the brain.

As also discussed by McConkey et al. (2013), it also does not appear that localisation of the parasite to a specific region of the brain explains behavioural changes. Vyas et al. (2007a) reported a subtle tropism for the amygdalar region, which is intriguing as this region is related to fear response. In contrast to Vyas et al. (2007a), however, most authors have reported that the parasite does not show a preference for any particular region of the brain (Berenreiterová et al., 2011). Afonso et al. (2012) found no evidence of an increase in brain cysts in amygdalar areas relative to other structures in the brain, although mice that had brain cysts in particular *combinations* of brain localisations were biased towards increased risk taking behaviour. They therefore hypothesised that the parasite may have evolved to manipulate functional circuits of the brain, rather than to manipulate a specific region. Most previous studies have not studied cyst localisation in this level of detail and therefore may have missed the subtle association between 'particular combinations of cyst locations' and change in behaviour.

5.2. Concluding remarks

The mechanism behind *T. gondii*-induced behavioural changes may involve direct and indirect effects of brain cysts, neurotransmitters, hormones and immune factors. Aside from investigating cyst load and localisation, most research into the proximate mechanism appears to be focused on changes in the dopaminergic system; this focus may be due to the types of behaviours that are considered to be affected by *T. gondii* and the general acceptance that behavioural changes are adaptive for the parasite, and so the mechanism is predicted to be *T. gondii*-specific (Vyas et al., 2007a; Webster et al., 2006). We suggest that while the proximate mechanism may be informed by the ultimate mechanism, and vice versa, neither should limit the other. To clarify, research concerning the proximate mechanism should not be limited by the hypothesis that behavioural changes are due to specific manipulation by the parasite, and research into potential mechanisms that do not involve specific manipulation by the parasite should be encouraged. As a mechanism of behavioural change, the modulation of immune response is an intriguing and promising avenue (Siegel and Zalcman, 2009) that deserves more attention.

6. WHERE SHOULD FUTURE RESEARCH FOCUS?

6.1. What factors influence the behavioural changes observed?

Attempts to elucidate the mechanism by which *T. gondii* induces behavioural changes may be hampered by the lack of consensus regarding what behaviours are affected by *T. gondii*. Insights from past studies suggest that differences in host species, host sex, *T. gondii* strain and dose, time post infection and the method used to measure behaviour may all help explain inconsistent results (Webster et al., 2013; Worth et al., 2013). A deeper understanding of the behavioural changes induced by *T. gondii* will provide a sound foundation for future research aimed at elucidating the mechanism(s) of action. To this end, future research should aim to clarify the influence of these experimental factors on behaviour, by utilising comparative experiments that allow the effect of particular factors to be determined. There is also a need to consider more carefully the reliability and validity of behavioural measurements, particularly through increasing the number of behavioural tests applied and developing more sophisticated analytical techniques.

6.2. The importance of individual differences

Although the goal of experimental infections is to determine whether *T. gondii* causes *changes* in behaviour, few studies to date have compared the behaviour of individuals both before and after *T. gondii* infection, instead electing to look at differences between two groups: infected and uninfected. Natural behavioural variations exist between individual rodents (Hogg and File, 1994; Kabbaj and Akil, 2001), which may explain the large levels of variation in control groups of past studies. It has been found that rats can be split into two groups based on natural variation in novelty-seeking behaviours; high responders are very responsive to novelty and have low anxiety-like behaviour, while low responders show low reactivity to novelty and have high anxiety-like behaviour (Kabbaj and Akil, 2001). Not only do high and low responders differ in their baseline gene expression in the brain, but they react to anxiety-provoking situations with different molecular responses, suggesting that anxiety-provoking stimuli are experienced differently by the two groups. Related to this, previous authors have pointed out that individual rats respond differently to cat odours (Kaushik et al., 2012; Vyas, 2013) and that some rats do not become 'attracted' to cat urine, even when infected with *T. gondii* (Vyas, 2013).

Such structured behavioural differences appear to be common in a very diverse range of animal species and may have important consequences for key ecological and evolutionary processes (Carter et al., 2013; Wolf and Weissing, 2012); the interaction of host personality, propensity for infection and subsequent behavioural manipulation by parasites is an area deserving of much more attention. In the context of examining behavioural manipulation of hosts by *T. gondii*, it would be useful to compare the behaviour of individuals at different time intervals before and after infection, with an uninfected group also included to control for repeated exposure to apparatus and natural changes in behaviour over time (e.g. due to ageing). A design of this type would enable both within-individual and between-individual variances and covariances in behavioural traits to be estimated (Dingemanse and Dochtermann, 2013).

6.3. Do behavioural changes actually enhance parasite fitness?

A key issue for future research regarding the ultimate mechanism is to determine, firstly, whether behavioural changes induced by *T. gondii* in rodents actually increase predation rate by cats and, secondly, the parameter space in which increased predation rate will enhance parasite fitness and therefore favour the evolution of behavioural manipulation. The first of these questions requires empirical testing, ideally in a natural or semi-natural situation. In the absence of a traditional predation experiment, it may be possible to test whether infected rodents are more likely to be caught by cats by comparing the prevalence of *T. gondii* in rodents caught by pet cats to the prevalence in the general population. The second question can be approached by extending the theoretical model of Lélu et al. (2013) to include the full range of transmission pathways for the parasite (Fig. 3.1). The value of such a model is not in directly determining the likelihood of behavioural manipulation being an adaptive trait (which would require additional information, such as the genetic architecture of the behavioural manipulation phenotype), but in indicating the conditions under which enhanced predation rate (of cats upon intermediate hosts) would increase the fitness (as represented, e.g. by the basic reproductive ratio) of the parasite. This then allows testable hypotheses to be developed concerning the relationship between behavioural manipulation and other parasite traits. For example, the analysis of Lélu et al. (2013) suggests that strains with a greater propensity for vertical transmission should be less likely to manipulate intermediate hosts to enhance predation, while strains that can transmit sexually from males to females should evolve higher rates of manipulation for female than for male intermediate hosts. Predictions such as this can be directly tested in experimental studies and their corroboration will greatly enhance the argument for behavioural manipulation of rodent behaviour by *T. gondii* being an adaptive trait of the parasite.

ACKNOWLEDGEMENTS

A. R. W. was in receipt of an Australian Postgraduate Award during this study. We would like to thank an anonymous reviewer for the comments that helped us to improve the clarity of the manuscript. A. R. W. would also like to thank Graeme Worth for being a willing participant in many a discussion regarding this topic. We would like to thank Mark Preston (Murdoch University) for creating the figure used in this chapter.

REFERENCES

Afonso, C., Paixão, V.B., Costa, R.M., 2012. Chronic *Toxoplasma* infection modifies the structure and the risk of host behaviour. PLoS One 7, e32489.

Arantes, T.P., Lopes, W.D.Z., Ferreira, R.M., Pieroni, J.S.P., Pinto, V.M.R., Sakamoto, C.A., da Costa, A.J., 2009. *Toxoplasma gondii*: evidence for the transmission by semen in dogs. Exp. Parasitol. 123, 190–194.

Araujo, F.G., William, D.M., Grumet, F.C., Remington, J.S., 1976. Strain-dependent differences in murine susceptibility to toxoplasma. Infect. Immun. 13, 1528–1530.

Berdoy, M., Webster, J.P., Macdonald, D.W., 1995. Parasite-altered behaviour: is the effect of *Toxoplasma gondii* on *Rattus norvegicus* specific? Parasitology 111, 403–409.

Berdoy, M., Webster, J.P., Macdonald, D.W., 2000. Fatal attraction in rats infected with *Toxoplasma gondii*. Proc. R. Soc. Lond. B 267, 1591–1594.

Berenreiterová, M., Flegr, J., Kuběna, A.A., Němec, P., 2011. The distribution of *Toxoplasma gondii* cysts in the brain of a mouse with latent toxoplasmosis: implications for the behavioural manipulation hypothesis. PLoS One 6, e23866.

Bollen, K.A., 1989. Structural Equations with Latent Variables. Wiley, New York.

Boyle, J.P., Saeij, J.P.J., Boothroyd, J.C., 2007. *Toxoplasma gondii*: inconsistent dissemination patterns following oral infection in mice. Exp. Parasitol. 116, 302–305.

Carter, A.J., Feeney, W.E., Marshall, H.H., Cowlishaw, G., Heinsohn, R., 2013. Animal personality: what are behavioural ecologists measuring? Biol. Rev. 88, 465–475.

Cézilly, F., Thomas, F., Médoc, V., Perrot-Minnot, M.-J., 2010. Host-manipulation by parasites with complex life cycles: adaptive or not? Trends Parasitol. 26, 311–317.

Chieffi, P.P., Aquino, R.T.R., Paschoalotti, M.A., Ribeiro, M.C.S.A., Nasello, A.G., 2009. Muscular strength decrease in *Rattus norvegicus* experimentally infected by *Toxocara canis*. Rev. Inst. Med. Trop. Sao Paulo 51, 73–75.

Darde, M.-L., Ajzenberg, D., Smith, J., 2007. Population structure and epidemiology of *Toxoplasma gondii*. In: Weiss, L.M., Kim, K. (Eds.), *Toxoplasma gondii*. The Model Apicomplexan—Perspectives and Methods. Academic Press, London.

Dass, S.A.H., Vasudevan, A., Dutta, D., Soh, L.J.T., Sapolsky, R.M., Vyas, A., 2012. Protozoan parasite *Toxoplasma gondii* manipulates mate choice in rats by enhancing attractiveness of males. PLoS One 6(11). e27229. http://dx.doi.org/10.1371/journal.pone.0027229.

Dingemanse, N.J., Dochtermann, N.A., 2013. Quantifying individual variation in behaviour: mixed-effect modelling approaches. J. Anim. Ecol. 82, 39–54.

Dubey, J.P., Jones, J.L., 2008. *Toxoplasma gondii* infection in humans and other animals in the United States. Int. J. Parasitol. 38, 1257–1278.

File, S.E., Hyde, J.R.G., 1978. Can social interaction be used to measure anxiety? Br. J. Pharmacol. 62, 19–24.

Flegr, J., 2007. Effects of *Toxoplasma* on human behaviour. Schizophr. Bull. 33, 757–760.

Flegr, J., 2013. Influence of latent *Toxoplasma* infection on human personality, physiology and morphology: pros and cons of the *Toxoplasma*-human model in studying the manipulation hypothesis. J. Exp. Biol. 216, 127–133.

Flegr, J., Kodym, P., Tolarová, V., 2000. Correlation of duration of latent *Toxoplasma gondii* infection with personality changes in women. Biol. Psychol. 53, 57–68.

Flegr, J., Havlíček, J., Kodym, P., Malý, M., Smahel, Z., 2002. Increased risk of traffic accidents in subjects with latent toxoplasmosis: a retrospective case–control study. BMC Infect. Dis. 2, 11.

Flegr, J., Preiss, M., Klose, J., Havlíček, J., Vitáková, M., Kodym, P., 2003. Decreased level of psychological factor novelty seeking and lower intelligence in men latently infected with the protozoan parasite *Toxoplasma gondii*. Dopamine, a missing link between schizophrenia and toxoplasmosis? Biol. Psychol. 63, 253–268.

Gatkowska, J., Wieczorek, M., Dziadek, B., Dzitko, K., Dlugonska, H., 2012. Behavioural changes in mice caused by *Toxoplasma gondii* invasion of brain. Parasitol. Res. 111, 53–58.

Gonzalez, L.E., Rojnik, B., Urrea, F., Urdaneta, H., Petrosino, P., Colasante, C., Pino, S., Hernandez, L., 2007. *Toxoplasma gondii* infection lower anxiety as measured by the plus-maze and social interaction test in rats: a behavioural analysis. Behav. Brain Res. 177, 70–79.

Goodwin, D., Hrubec, T.C., Klein, B.G., Strobl, J.S., Werre, S.R., Han, Q., Zajac, A.M., Lindsay, D.S., 2012. Congenital infection of mice with *Toxoplasma gondii* induces minimal change in behavior and no change in neurotransmitter concentrations. J. Parasitol. 98, 706–712.

Gulinello, M., Acquarone, M., Kim, J.H., Spray, D.C., Barbosa, H.S., Sellers, R., Tanowitz, H.B., Weiss, L.M., 2010. Acquired infection with *Toxoplasma gondii* in adult mice results in sensorimotor deficits but normal cognitive behaviour despite widespread brain pathology. Microbes Infect. 12, 528–537.

Havlíček, J., Gašová, Z., Smith, A.P., Zvára, K., Flegr, J., 2001. Decrease of psychomotor performance in subjects with latent 'asymptomatic' toxoplasmosis. Parasitology 122, 515–520.

Hay, J., Aitken, P.P., Hutchison, W.M., Graham, D.I., 1983a. The effect of congenital and adult-acquired *Toxoplasma* infections on the motor performance of mice. Ann. Trop. Med. Parasitol. 77, 261–277.

Hay, J., Hutchison, W.M., Aitken, P.P., Graham, D.I., 1983b. The effect of congenital and adult-acquired *Toxoplasma* infections on activity and responsiveness to novel stimulation in mice. Ann. Trop. Med. Parasitol. 77, 483–495.

Hay, J., Aitken, P.P., Hair, D.M., Hutchison, W.M., Graham, D.I., 1984. The effect of congenital *Toxoplasma* infection on mouse activity and relative preference for exposed areas over a series of trials. Ann. Trop. Med. Parasitol. 78, 611–618.

Hodkova, H., Kodym, P., Flegr, J., 2007. Poorer results of mice with latent toxoplasmosis in learning tests: impaired learning processes or the novelty discrimination mechanism? Parasitology 134, 1329–1337.

Hogg, S., File, S.E., 1994. Responder and nonresponders to cat odor do not differ in other tests of anxiety. Pharmacol. Biochem. Behav. 49, 219–222.

Hrdá, S., Votýpka, J., Kodym, P., Flegr, J., 2000. Transient nature of *Toxoplasma gondii*-induced behavioral changes in mice. J. Parasitol. 86, 657–663.

Hughes, R.N., 2004. The value of spontaneous alternation behavior (SAB) as a test of retention in pharmacological investigations of memory. Neurosci. Biobehav. Rev. 28, 497–505.

Hutchison, W.M., Aitken, P.P., Wells, B.W.P., 1980a. Chronic *Toxoplasma* infections and familiarity-novelty discrimination in the mouse. Ann. Trop. Med. Parasitol. 74, 145–150.

Hutchison, W.M., Aitken, P.P., Wells, B.W.P., 1980b. Chronic *Toxoplasma* infections and motor performance in the mouse. Ann. Trop. Med. Parasitol. 74, 507–510.

Hutchison, W.M., Bradley, M., Cheyne, W.M., Wells, B.W.P., Hay, J., 1980c. Behavioural abnormalities in *Toxoplasma*-infected mice. Ann. Trop. Med. Parasitol. 74, 337–345.

Innes, E.A., 1997. Toxoplasmosis: comparative species susceptibility and host immune response. Comp. Immunol. Microbiol. Infect. Dis. 20, 131–138.

Johnson, A., 1997. Speculation on possible life cycle for the clonal lineages in the genus *Toxoplasma*. Parasitol. Today 13, 393–397.

Kabbaj, M., Akil, H., 2001. Individual differences in novelty-seeking behavior in rats: a c-*fos* study. Neuroscience 106, 535–545.

Kannan, G., Moldovan, K., Xiao, J., Yolken, R.H., Jones-Brando, L., Pletnikov, M.V., 2010. *Toxoplasma gondii* strain-dependent effects on mouse behaviour. Folia Parasitol. 57, 151–155.

Kaushik, M., Lamberton, P.H.L., Webster, J.P., 2012. The role of parasites and pathogens in influencing generalised anxiety and predation-related fear in the mammalian central nervous system. Horm. Behav. 62, 191–201.

Kempf, M.-C., Cesbron-Delauw, M.-F., Deslee, D., Groß, U., Herrmann, T., Sutton, P., 1999. Different manifestations of *Toxoplasma gondii* infection in F344 and LEW rats. Med. Microbiol. Immunol. 187, 137–142.

Lagrue, C., Poulin, R., 2010. Manipulative parasites in the world of veterinary science: implications for epidemiology and pathology. Vet. J. 184, 9–13.

Lamberton, P.H.L., Donnelly, C.A., Webster, J.P., 2008. Specificity of the *Toxoplasma gondii*-altered behaviour to definitive versus non-definitive host predation risk. Parasitology 135, 1143–1150.

Lélu, M., Langlais, M., Poulle, M.-L., Gilot-Fromont, E., Gandon, S., 2013. When should a trophically and vertically transmitted parasite manipulate its intermediate host? The case of *Toxoplasma gondii*. Proc. R. Soc. Lond. B 280, 20131143.

Mcconkey, G.A., Martin, H.L., Bristow, G.C., Webster, J.P., 2013. *Toxoplasma gondii* infection and behaviour—location, location, location? J. Exp. Biol. 216, 113–119.

Poulin, R., 2010. Parasite manipulation of host behaviour: an update and frequently asked questions. In: Brockmann, H.J. (Ed.), Advances in the Study of Behaviour. Academic Press, Burlington.

Prandovszky, E., Gaskell, E.A., Martin, H., Dubey, J.P., Webster, J.P., Mcconkey, G.A., 2011. The neurotropic parasite *Toxoplasma gondii* increases dopamine metabolism. PLoS One 6, e23866.

Remington, J.S., Cavanaugh, E.N., 1965. Isolation of the encysted form of *Toxoplasma gondii* from human skeletal muscle and brain. N. Engl. J. Med. 273, 1308–1310.

Roberts, C.W., Cruickshank, S.M., Alexander, J., 1995. Sex-determined resistance to *Toxoplasma gondii* is associated with temporal differences in cytokine production. Infect. Immun. 63, 2549–2555.

Saeij, J.P.J., Boyle, J.P., Boothroyd, J.C., 2005. Differences among the three major strains of *Toxoplasma gondii* and their specific interactions with the infected host. Trends Parasitol. 21, 476–481.

Seppälä, O., Jokela, J., 2008. Host manipulation as a parasite transmission strategy when manipulation is exploited by non-host predators. Biol. Lett. 4, 663–666.

Siegel, A., Zalcman, S.S. (Eds.), 2009. The Neuroimmunological Basis of Behavior and Mental Disorders, Springer, USA.

Skallová, A., Kodym, P., Frynta, D., Flegr, J., 2006. The role of dopamine in *Toxoplasma*-induced behavioural alterations in mice: an ethological and ethopharmacological study. Parasitology 133, 525–535.

Taboas, W., Mckay, D., 2012. Does *Toxoplasma gondii* play a role in obsessive-compulsive disorder? Psychiatry Res. 198, 176–177.

Tomer, A., Pugesek, P.H., 2003. Guidelines for the implementation and publication of structural equation models. In: Pugesek, P.H., Tomer, A., Von Eye, A. (Eds.), Structural Equation Modelling: Applications in Ecological and Evolutionary Biology. Cambridge University Press, Cambridge.

Torrey, E.F., Bartko, J.J., Lun, Z.-R., Yolken, R.H., 2007. Antibodies to *Toxoplasma gondii* in patients with schizophrenia: a meta-analysis. Schizophr. Bull. 33, 729–736.

Vervaeke, M., Davis, S., Leirs, H., Verhagen, R., 2006. Implications of increased susceptibility to predation for managing the sylvatic cycle of *Echinococcus multilocularis*. Parasitology 132, 893–901.

Vyas, A., 2013. Parasite-augmented mate choice and reduction in innate fear in rats infected by *Toxoplasma gondii*. J. Exp. Biol. 216, 120–126.

Vyas, A., Sapolsky, R.M., 2010. Manipulation of host behaviour by *Toxoplasma gondii*: what is the minimum and proposed proximate mechanism should explain? Folia Parasitol. 57, 88–94.

Vyas, A., Kim, S.-K., Sapolsky, R.M., 2007a. The effects of toxoplasma infection on rodent behaviour are dependent on dose of the stimulus. Neuroscience 148, 342–348.

Vyas, A., Kim, S., Giacomini, N., Boothroyd, J.C., Sapolsky, R.M., 2007b. Behavioural changes induced by *Toxoplasma gondii* infection of rodents are highly specific to aversion of cat odors. Proc. Natl. Acad. Sci. U.S.A. 104, 6442–6447.

Walf, A.A., Frye, C.A., 2007. The use of the elevated plus maze as an assay of anxiety-related behaviour in rodents. Nat. Protoc. 2, 322–328.

Walsh, R.N., Cummins, R.A., 1976. The open-field test: a critical review. Psychol. Bull. 83, 482–504.

Webster, J.P., 1994. The effect of *Toxoplasma gondii* and other parasites on activity levels in wild and hybrid *Rattus norvegicus*. Parasitology 109, 583–589.

Webster, J.P., 2001. Rats, cats, people and parasites: the impact of latent toxoplasmosis on behaviour. Microbes Infect. 3, 1037–1045.

Webster, J.P., 2007. The effect of *Toxoplasma gondii* on animal behaviour: playing cat and mouse. Schizophr. Bull. 33, 752–756.

Webster, J.P., Mcconkey, G.A., 2010. *Toxoplasma gondii*-altered host behaviour: clues as to a mechanism of action. Folia Parasitol. 57, 95–104.

Webster, J.P., Brunton, C.F.A., Macdonald, D.W., 1994. Effect of *Toxoplasma gondii* upon neophobic behaviour in wild brown rats, *Rattus norvegicus*. Parasitology 109, 37–43.

Webster, J.P., Lamberton, P.H.L., Donnelly, C.A., Torrey, E.F., 2006. Parasites as causative agents of human affective disorders? The impact of anti-psychotic, mood-stabilizer and anti-parasite medication on *Toxoplasma gondii*'s ability to alter host behaviour. Proc. R. Soc. Lond. B 273, 1023–1030.

Webster, J.P., Kaushik, M., Bristow, G.C., Mcconkey, G.A., 2013. *Toxoplasma gondii* infection, from predation to schizophrenia: can animal behaviour help us understand human behaviour? J. Exp. Biol. 216, 99–112.

Witting, P.-A., 1979. Learning capacity and memory of normal and *Toxoplasma*-infected laboratory rats and mice. Parasitol. Res. 61, 29–51.

Wolf, M., Weissing, F.J., 2012. Animal personalities: consequences for ecology and evolution. Trends Ecol. Evol. 27, 452–461.

Worth, A.R., Lymbery, A.J., Thompson, R.C.A., 2013. Adaptive host manipulation by *Toxoplasma gondii*: fact or fiction? Trends Parasitol. 29, 150–155.

Xiao, J., Kannan, G., Jones-Brando, L., Brannock, C., Krasnova, I.N., Cadet, J.L., Pletnikov, M., Yolken, R.H., 2012. Sex-specific changes in gene expression and behavior induced by chronic Toxoplasma infection in mice. Neuroscience 206, 39–48.

INDEX

Note: Page numbers followed by "*f*" indicate figures and "*t*" indicate tables.

Q

R

S

T

CONTENTS OF VOLUMES IN THIS SERIES

Volume 41

Volume 42

Volume 43

Volume 44

Volume 45

Volume 46

Volume 47

Volume 48

Volume 49

Volume 50

Volume 51

Volume 52

Volume 53

Volume 54

Volume 55

Volume 61

Volume 62

Volume 63

Volume 64

Volume 65

Volume 66

Volume 67

Volume 68

Volume 69

Volume 70

Volume 71

Volume 72

Volume 75

Volume 76

Volume 77

Volume 78

Volume 79

Volume 80

Volume 81

Volume 82

CPI Antony Rowe
Eastbourne, UK
June 22, 2014